T0171441

the match

COMPLETE STRANGERS,
A MIRACLE FACE TRANSPLANT,
TWO LIVES TRANSFORMED

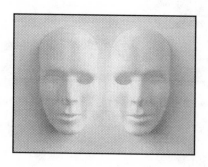

susan whitman helfgot

with william novak

Simon & Schuster

New York   London   Toronto   Sydney

Simon & Schuster
1230 Avenue of the Americas
New York, NY 10020

First Simon & Schuster hardcover edition October 2010

SIMON & SCHUSTER and colophon are registered trademarks
of Simon & Schuster, Inc.

For information about special discounts for bulk purchases,
please contact Simon & Schuster Special Sales at
1-866-506-1949 or business@simonandschuster.com.

The Simon & Schuster Speakers Bureau can bring authors
to your live event. For more information or to book an event,
contact the Simon & Schuster Speakers Bureau at
1-866-248-3049 or visit our website at www.simonspeakers.com.

Designed by Nancy Singer

Manufactured in the United States of America

10   9   8   7   6   5   4   3   2   1

Library of Congress Cataloging-in-Publication Data
Helfgot, Susan Whitman.
    The match : complete strangers, a miracle face transplant, two lives
transformed / Susan Whitman Helfgot, with William Novak.
        p. cm.
    1. Maki, James—Health. 2. Helfgot, Joseph—Health. 3. Face—Transplanta-
tion—Biography. I. Novak, William. II. Title.
        RD523.M35 2010
        617.5'205920922—dc22

2010022476

ISBN 978-1-4391-9549-9

for joseph

He had a face like a blessing.

—Cervantes

the match

chapter one

For a dying man, it is not a difficult decision because he knows he is at the end. If a lion chases you to the bank of a river filled with crocodiles, you will leap into the water, convinced you have a chance to swim to the other side.

—Dr. Christaan Barnard

Monday, April 6, 2009, late morning.
Brigham and Women's Hospital, Boston.

a calculated quiet has fallen over the normally frenetic and noisy cardiac intensive care unit. Nurses stand in groups of two or three, speaking quietly. Others attend to their patients' life-or-death concerns in slow and measured movements out of respect for what has just happened.

Death has come to the floor. Not just any death, although all deaths are tragic in the cardiac surgical ICU, but the death of a patient who has become a fixture here, and for many of the staff, a friend. Joseph Helfgot had battled heart disease for more than a decade, and for the past two years he was cared for by the people who now stand around in stunned silence.

He came so damn close, inching up on the list of patients waiting for a new heart. Only a third of them ever get there, and he was one of the lucky ones. Just two nights ago the New England Organ Bank finally matched the heart from a man who had just died. Word spread quickly. Helfgot was watching late-night TV while his wife slept. The phone woke her up. "Mrs. Helfgot, it's Dr. Lewis. I think we found a heart. Don't rush, take your time, but start putting things together."

"Joseph, we have a heart!"

Ignoring the doctor's advice, they rushed to the hospital, barely taking time to say goodbye to their two boys, who were camped out in the family room, half asleep, ready for bed.

"Bye, Dad. We'll see you after the operation."

At the hospital e-mails flew around the floor. Off-duty staff were copied: Joe's getting a heart! That was Saturday night. Two days later elation has turned to grief.

Earlier that morning.

Dr. Jim Rawn, the surgeon who has orchestrated Helfgot's day-to-day care during his frequent stays in the ICU, steps into the surgical unit. It is barely 6 a.m., but the place is jumping. Two new hearts came in over the weekend, Helfgot's and another one.

Dr. Rawn likes Helfgot, a market research executive who works in the movie business. He knows he has broken a cardinal rule: Don't get too close. But sometimes a heart patient pierces through the cloak of aloofness that intensive care physicians wear like armor. Hollywood Joe, as the nurses call him, is one of them. Rawn has learned the hard way that he shouldn't become too attached. Although the Brigham's cardiac unit is one of the finest in the country, not every transplant patient who comes in here will walk out the door. Better to check your emotions before you come to work, because it hurts too much when you get close.

But sometimes you can't resist, and Joseph Helfgot can shatter the toughest of façades. When he's not knocking on death's door, it's hard to believe he's a patient at all. In the ICU his bed is usually littered with half a dozen movie scripts. Piles of yellow legal pads filled with notes cover the top of his hospital tray, and a second tray on the other side of his bed holds his laptop and his BlackBerry. He has set up shop here. The nurses call it his bed-quarters, a term his employees have been using for years. Even before he got sick Helfgot loved to work from bed.

In the hospital he regales anyone who cares to listen, as well as a few who don't, with behind-the-scenes stories about the "real" Hollywood. It isn't uncommon to find a heart specialist with a gaggle of medical students crowding around Helfgot, who is propped up in bed with a movie booming loudly on his laptop. "So I'm watching *Public Enemy*. You know, Jimmy Cagney plays a gangster? They're doing a kind of remake, with Johnny Depp as John Dillinger. So which do you like better? The scene over here, where Cagney shoots the guy?" The students lean in. "Or this one, where he kisses the girl?" He fast-forwards the movie as they stare blankly at the screen. "Christ, how old are you guys anyway? Do you even know who James Cagney *is*? How about Jean Harlow? You know, the blonde? Crap, never mind."

Some of the nurses adore him—mostly those he hasn't driven half-crazy with his perpetual list of demands. With a few of them he has the kind of relationship they have with their hairdresser. He can also be exasperating.

"I'm sorry, Judy, but I need this ice-cold, please."

"Mr. Helfgot, we're talking about liquid potassium. It's medicine, not a cocktail. We don't serve it on the rocks."

"Just bring me some ice, please," he says, flashing a petulant smile.

Judy shakes her head. The first time she met him, he asked if she was married.

"No," she said, busily attending to an occluded IV. "Put your arm out and try to hold still, will you?"

"Why not?" he asked, inches from her face as she checked his line, looking for the chink in the tubing.

She stared back at him blankly, thinking, Boy, *that's* personal. But then she heard herself saying, "Good question. Why the hell am I not married?"

"You'll find somebody," he told her. "You're pretty."

Helfgot's wife was there, and he turned to her and said, "Susan, do we know anyone for her?" To the nurse's embarrassment, which conveniently masked how much she was enjoying this conversation, Joseph and Susan began ticking off names of the single men they knew, and why this one or that one would be suitable or not. A year and a few dates later, some with Helfgot's bachelor friends, Judy was still single.

"I'll get your ice, but you have to drink it all at once, and not over the course of the next two weeks. Your K is so low you're going to crash."

"Okay. I promise."

Dr. Rawn steps quietly into Helfgot's room. This morning, right after the transplant, there aren't any scripts on the bed. Just Helfgot. He hasn't woken up, but it was a long surgery.

"He's not waking up," the nurse says. The nurse is worried, but he's trying not to show it.

Rawn stands over the bed and does his own quick check. He lifts Helfgot's eyelids and examines his pupils, which are dilated. He takes the flashlight from the wall and shines it right into Helfgot's eyes. Nothing happens, no contraction at all. *Shit.* He calls for a CT scan and the nurse picks up the phone.

An hour later, after the scan, Rawn hovers around the computer screen at the physician's desk outside Helfgot's room, anxiously waiting for the results to upload. He nods at Susan, who has just arrived.

"Not awake yet?" she asks the nurse.

"You know how long it takes for Mr. Helfgot to wake up from surgery."

She does indeed. This is his fourth surgery in a year and a half. Although this is the big one, the one they have long hoped was coming, she is numb from the constant fear of his death. She hasn't had any real sleep in more than a year. This whole medical adventure has been a prolonged road trip through hell.

"Wake up, Joseph!" she shouts in his ear. Sometimes hearing a familiar voice does the trick. She lifts his eyelids and then shakes him a few times. An angry vitals monitor picks up the disturbance and sends out an alarm. Susan reconnects an EKG lead and instinctively pushes the reset button on the monitor high above her head. After a year of bringing her husband in and out of the ICU she knows her way around the machinery, although she also knows that she shouldn't be touching the equipment. She turns to the nurse and says, "You didn't see that."

Dr. Rawn, waiting at the computer, is watching through the glass wall of the room and thinking that Susan would make a great nurse. But why is it taking so damn long for these results? Finally a lateral view of Helfgot's head appears on the screen before him. He stares at it, not wanting to believe what he is seeing. A third of the right side of the brain is in darkened shadow, and some of the left side as well. God damn it! Massive ischemia. The brain architecture is gone. Clots must have traveled up during surgery, closing off the blood supply to Helfgot's head.

Rawn spots Greg Couper, Helfgot's heart surgeon, and motions for him to come over. "Greg, take a look."

Couper pulls off his glasses and peers at the screen. He tries not to show any expression as his stomach sinks down to his toes. During the surgery he had found a clot in Helfgot's aorta, which was not a good sign. He mentioned it to Susan when they spoke after the operation. He wasn't sure she grasped the significance of the information, but she had been with Helfgot's son, and it wasn't the time or the place to raise fears. He had found clots before; they don't always mean a bad ending, but often they do.

"Have you spoken to his wife?" he asks Rawn.

"No."

A few nurses are looking over in their direction with *We know something's up* expressions.

"The heart is doing well?" Couper asks.

"Banging away." With a respirator still bringing air into Helfgot's lungs, the heart can continue beating.

Stupid! Couper thinks. This whole thing is just too stupid. After all that work to keep him alive. After the artificial heart pump he'd implanted a year and a half ago to keep him going while he waited for a heart, which almost killed the poor guy from so many complications. But Joseph Helfgot was a fighter. He was determined to make it to his son's bar mitzvah, which he talked about all the time. Well, at least he got there. And now, a few months later, he dies getting the transplant? Damn it.

Couper is tired. He performed a second heart transplant only hours after scrubbing out of Helfgot's surgery, and he hasn't been to bed in over twenty-four hours. Transplants are a strange and unpredictable business. Sometimes it's quiet for days on end, and at other times it feels like sheer insanity, with hearts flying in like planes over LaGuardia. And you never know when one of your patients is going to die.

Like right now.

A former college wrestler, Couper had an instinct to hang on, to cling to something to keep his patient off the mat. He stares at the composite X-ray of Helfgot's brain, and the image makes him angry. He wants to pick up the monitor and hurl it through the nearest window.

"Okay," he says, "let's get the heart checked out. Maybe someone else can use it."

"We're doing one more scan just to be sure," Rawn tells him, but he knows the test is pointless. This movie is over. Couper goes off to check on the other transplant patient who has come up from surgery.

Dr. Rawn prepares himself to walk into the room and inform the widow, who doesn't yet know she's a widow. For a long time he was never sure whether to call her Mrs. Helfgot or Mrs. Whitman. Finally it just became Susan. Now it's Susan the widow.

"So, what's going on?" she asks.

He has always been candid with her. She has a knack for medicine and has been a tireless advocate for her husband.

"He should be awake by now. We did a scan. It doesn't look good."

She suddenly knows that this is the worst moment of her life.

"Dr. Couper found a clot in the aorta during surgery," she says. "He's stroked out, hasn't he?"

"We're doing another scan."

"Why?"

"We just are. There might be something . . ."

They stand there for a moment. "It's okay," she says. "I get it."

He nods, unable to speak.

"So that's that, isn't it?"

He nods again.

"I'm really sorry, Jim," she says as she reaches out and touches his sleeve.

I had thought again and again about how not to fall apart when it finally happened, because I knew it would. The odds were always against us. It soothed me to go off in my mind and practice his death. It made me feel safer, more prepared.

And now Jim Rawn is looking at me, his eyes telling me Joseph is dead. Now there is no inside of me—only outside. All my rehearsing has paid off. The inside will come back. Later.

Just like that, it's over. No more ideas to try, no meds to tweak, no specialists to consult, no waiting for a damn heart. Just like that.

Now there will be crying children, phone calls to make, a casket to buy, plane tickets and food to arrange—all the unrelenting busy-ness that is the sole blessing attached to death.

On another floor of the hospital Esther Charves, a family coordinator with the New England Organ Bank, has just finished a case. On the elevator she recognizes one of the chaplains, a southern woman who has just left the ICU and who knows Helfgot well. Among other topics the chaplain and Helfgot liked to argue about how to make good barbecue. "Joseph," she would admonish him in her thick drawl, "you shouldn't be eating that stuff at all. Too much salt." But most of the time Joseph ate what he wanted.

The chaplain tried to go easy on Helfgot, for although his dietary habits could be abominable, she admired the way he never complained about his bad luck—and he had a lot of it. In her line of work the standard question is "Why has God done this to me?" Helfgot never asked. Maybe he had an answer, or maybe he knew that it didn't help to wonder.

She knew patients who just lay there, like inmates on death row, cursing their bad luck. Helfgot wasn't one of them. Sometimes he would get up and stroll around the ICU, trying to cheer everybody up. One day he handed out sushi that Susan had smuggled in.

"Try this urchin. It's *amazing.*"

"Urchin? Um, no thanks."

"No, really, you *have* to try this. It's unbelievable."

"I'd love to," one patient told him, "but the doctors have put me on an urchin-free diet."

"Esther," the chaplain says, "you may have a case up on six. Heart transplant gone bad."

"How old?" Young people dying are the worst.

"Late fifties, maybe sixty."

Esther gets off on the sixth floor and enters the unit, where a few nurses are crying. She has witnessed this before, but not often. Jim Rawn greets her and fills her in, adding, "You may want the heart too."

A retransplant of a newly transplanted heart? "I've never seen that," she says.

"I haven't either, and I've been here ten years. But we think it's viable, and Dr. Couper wants to take a look."

She glances into Helfgot's room. Close to a dozen people, mostly hospital staffers in scrubs and white coats, surround a small woman Esther can barely see. She must be the widow.

Esther is optimistic. A family that has just received an organ, even if it went badly, will likely reciprocate if they have the chance. She'll come back in a little while, when things are more settled.

Her cell phone rings. It's Chris Curran, the head coordinator at the organ bank. "Are you on the cardiac unit?" he asks her.

"I just got here."

"They say the heart's working fine."

"That's what Jim Rawn just told me."

"We may have a match in the Midwest," says Curran. "Their team is checking it out."

On Saturday, when the organ bank matched Helfgot with a heart, his name was moved from Waiting to Transplanted in the national database. Curran has never met Helfgot, but for years he has watched his name moving up and down the list. It's always satisfying when a patient finally makes it to Transplanted. Now they will have to move him over to Deceased. He looks at the information next to Helfgot's name: type O, age sixty.

"Esther? He could be the one we've been waiting for. We'll check it out on our end."

"Are you sure? The heart's a big enough deal. And I don't remember any of us ever asking for a retransplanted heart. I haven't even met Mrs. Helfgot yet. It's way too soon to go there."

"But do you think it's possible?"

"Chris, I just got here. I'll call you in a little while. I need to walk."

Esther drops the cell phone into her pocket. It's too soon, but her heart is beating faster anyway. They've been on the lookout for quite a while. She peers back into the room. The widow isn't there; she must have left during the phone call. A few doctors are at the patient's bedside. Esther recognizes one of them. He looks upset.

A few days ago the organ bank team thought they had their special donor, but it didn't pan out. This one could work. Helfgot's family might be open to it. Esther walks around the unit, past the room of the other man who was transplanted over the weekend. She pulls out her phone.

"Chris? I'll speak with her about the heart when she comes back—but only if you're sure there is someone who will take it. And if that works out, I'll talk with her about the other thing."

Esther wonders when Mrs. Helfgot will be back. Maybe she'll be the one, the one who will say "Yes, you can have my husband's face."

chapter two

Monday, April 6, 2009.
A veterans' home in central Massachusetts.

James Maki stares at his computer screen. A few doors away an old church at the foot of the hill peals off the familiar Westminster chimes. As the hour strikes, Maki stops to count. Eleven o'clock. The toolbar in front of him gives the time, but it's too tiny for him to see. His right eyelid is stitched shut. The nerve that opens and closes it was destroyed four years ago in the accident.

One of his housemates is cooking something downstairs. It smells like grilled cheese, although it's a little early for lunch. He'll go down in an hour or so. On the screen three cards show up: a ten, an eight, and a three. He's holding a queen and a seven. He folds.

This hasn't been one of his better days, but he keeps playing. Texas Hold'em eats up a lot of time, and time is all he has. The screen clears and he pulls an unsuited deuce and seven—the worst pair you can draw.

A car will be coming at five to take the residents to Applebee's for dinner, but Maki won't be joining them. Maybe he can talk the woman who runs the house into bringing something back for him, like one of those sundae shooters in a cup. She knows about his sweet tooth.

He'd love to go with them, but it's not a good idea. People don't want to look at a big crater where a person's face ought to be, especially when they're out to dinner. And because the top of his mouth is missing, half of what he tries to eat ends up on his lap. His housemates, who are also middle-aged Vietnam vets, have their own problems. Some are missing arms or legs, or are using wheelchairs, or are on oxygen, and they draw enough stares without him tagging along. Jim Maki draws more than stares. People have been known to scream when they see him. No thank you. Maybe he'll have macaroni and cheese tonight. A pair of jacks? Now that's more like it.

He keeps clicking away. Up three hands, down the next two. He plays so much that he barely pays attention and shifts over to the game show on his flat-screen TV. He's got a pretty nice setup, all things considered. True, the room is small and the floor is linoleum. Wall-to-wall carpeting would be nice. But the TV and the computer are top of the line. His brother, John, who lives in Seattle, bought the computer when Jim arrived here a year ago from the rehab place in New Bedford.

John hasn't been back to visit for a while. He hates to fly because he gets claustrophobic on planes. But he came several times when Jim was in rehab to make sure his younger brother was well cared for. Because Jim is a sharp dresser, John bought him some beautiful clothes. Back in the day Jim used to shop at Louis, Boston's premium men's clothier.

Jim's wife, Cynthia, from whom he is separated, advised John to return the clothes to wherever he bought them because they soon would be stolen. She was right. Most of them disappeared within a few days.

He looks out the window, past the cheap white polyester curtain that hangs down the middle in a loose knot. It's a miserable day, but even with the rain it's good to look outside. Someone said this morning that Opening Day might be rained out. Downstairs the phone is ringing, and a few seconds later he hears someone's name called out from the foot of the stairs. So the call isn't for him. But someday it could be. Someday Dr. Pomahac could be calling him.

He can't remember much about the day he ended up at the Ruggles subway station in Boston, where he fell onto the tracks and lost his face.

He was living in a halfway house in Malden and sticking to his methadone treatment. But he was still using drugs whenever he could, trying not to get caught and be thrown out. He was strung out that night and somehow fell onto the tracks. He lost the use of his right hand, which dangles from his arm. He can live with that, but his face hit the third rail dead on. Thank God he can't remember any of it.

Dr. Pomahac operated on him ten times, until there was nothing more he could do. Then Jim saw him on TV, talking about face transplants. He sat transfixed as Pomahac explained how this procedure, which was still mostly theoretical, might eventually help horribly disfigured patients.

"I saw you on TV," he said at his next appointment. "Do you think I could have a face transplant?"

"We're working on it, Mr. Maki." The young Czech surgeon knew that if anybody needed such an extreme intervention, it was Maki. "But we're not quite there yet."

Then one day, seemingly out of nowhere, Dr. Pomahac said, "If we could give you a new face, which carries some grave risks, including the possibility of getting cancer from the rejection drugs you would have to take for the rest of your life, would you still want to do it?"

Maki didn't hesitate. "Absolutely."

He'd had so many tests, and with each one he worried that they'd find something wrong, that something would change Dr. Pomahac's mind and rule him out. So far that hasn't happened. He's still hanging on. The doctor has told him he will get a face as soon as they find a donor. But who knows when that will happen. Or if it ever will.

It's hard not to think about it every time the phone rings. Jim knows this could take a long time, but the waiting has been hard. He feels awful knowing that someone has to die for him to get his face. At the hospital he has talked about it with the psychiatrist. Dr. Pomahac says that he won't look anything like the donor, which is a relief. But the whole thing is still pretty weird.

In the distance he can hear what sounds like one of his housemates talking on the phone to one of his children. He instinctively looks over at a picture of Jessica. Beautiful Jessie, with her huge almond eyes and exotic cheeks, a perfect blend of him and Cindy. She stares back at him with a big smile under a straw hat. Eighteen, with creamy white skin and not a wrinkle on her face, she is a perfect flower. Soon she hopes to go to Korea to teach English. Her grandmother would be so proud.

His thoughts turn to his mother, and he wishes she were here to comfort him. Mary Maki was a sweet and gentle woman who never turned her back on him, even after all the hurt he put her through. Lately he's been thinking about her a lot. He wasn't sober when she died more than ten years ago, and it is only recently, now that his mind is finally clear, that he feels the crushing grief.

Fifty-two years earlier. Winter 1957. Seattle.
Mary Maki looks out her kitchen window at the swirling clumps of sticky white snowflakes. They have been falling steadily for over an hour, and she can barely see across her yard. Seattle receives its share of snow each winter, but the mild sea breezes blowing in from Puget Sound usually melt it pretty fast. This morning looks like one of those times that the snow will stick. Maybe they will have to cancel school.

She runs her spatula across the bottom of a skillet, gently prodding the eggs into a fluffy mass. Her husband left for the university at the crack of dawn, off to another early breakfast meeting of some committee or another. It's a wonder he has any time to teach. If this snow keeps up, he may come home early for a change.

"Boys," she calls out, "your breakfast is ready."

John Jr. and Jim, who are nine and seven, take their seats at the table for toast and jam, eggs, and fruit. There are no boxes of Cheerios or, heaven forbid, that Kellogg's Special K everyone is talking about. She pours milk into plastic cups and sits down next to her children.

"Mom, you think there's gonna be school today?"

"I don't know, Jim. Eat your breakfast."

The phone rings. It's the wife of one of her husband's colleagues, who tells Mary that school is closed.

"Yay!" the boys shout.

She smiles at her two adopted sons. "Sit down, boys. Finish your breakfast."

"Mom, can we go sledding?" Jim asks as he shovels eggs into his mouth.

She considers Jim's face as he swallows the eggs with a large gulp of milk, his high, wide cheekbones peeking from the rim of the cup. Where did he get those cheeks? The older he gets, the more certain she is that the half of him that isn't Japanese is something other than white American. There's no doubt about the half-Japanese part; his smoky black eyes under almond eyelids attest to that. But for the past year or two she has wondered about the other half of his ancestry. Maybe he's part Native American?

"I think you should stay in the yard today."

"*Please?*" he begs.

His brother joins in the appeal: "*Please, Mom?*"

They are the only Japanese family in the neighborhood. The people on their street had a meeting when they learned that John and Mary Maki had put an offer on the house, which is slightly grand by the modest standards of this blue-collar area. A few residents tried to think up ways to keep the Japs from moving in. Pearl Harbor wasn't that long ago, but this is about more than the war. John and Mary find it hard to imagine that German Americans would be treated this way.

Professor John Maki, who is known as Jack, is a noted Asian scholar at the University of Washington who worked for the U.S. government during the war. He may be Japanese, but he is educated and well regarded by Seattle's academic community. On 125th Street, just a block from the lake, the bigotry is mostly tucked away behind closed doors. But it's definitely there.

It was easier when the children were little. Mary kept them in the yard and invited children from families they knew to come over and play. But now the boys are older, and a wider world is out there, waiting for them. She watched sadly one day as they stood at the edge of the yard while a group of boys ran down the street in cowboy hats and holsters with toy guns. Jack was adamant that their boys should stay in their yard.

Just after they were married Jack and Mary had experienced a far more shocking form of prejudice. Soon after Pearl Harbor, like other Japanese Americans on the West Coast, they were forced into an internment camp. The Makis were fortunate: their stay ended after a month, when Jack was summoned to Washington, D.C. One day you're a security risk; the next day you're working for a government intelligence agency. Most of their relatives paid a higher price for their ancestry.

"I'd rather you boys played in the yard," Mary tells them. "Maybe you and your friends can build a snowman."

"Mom, that's *boring*. We want to go sledding. We've got the best hill in the whole neighborhood." Jim is already out of his seat and wiping his milky mustache with his sleeve. Before she can object, he grabs his boots and coat and dashes outside, sending a flurry of white powder into the room. It was true: kids came from several blocks away to sled down their street.

"John, don't let him out of your sight. And don't go past the corner. Do you understand me?"

"Yes, Mom. You're the best mother there ever was!"

As Mary scrapes the half-eaten breakfast plates into the sink, she is already planning lunch. Maybe Jack will come home early. A nourishing soup? She opens the door of the fridge and starts pulling out vegetables. While other housewives may wonder whether to use light cream or half-and-half when they prepare Campbell's condensed tomato soup, Mary makes everything from scratch. There are no frozen TV dinners in Mary Maki's shopping cart.

Her father, Ichi, was Isei, a Japanese-born immigrant. A Christian farmer, he stuck to traditional ways, even insisting that Japanese be spoken in his American farmhouse. He did well and eventually installed a train track on the farm to ship potatoes and rhubarb to the railroad station, where his produce was added to the freight traveling east. He and his wife, Rin, raised five children on the farm.

Often, when Mary bends down to smell a flower or pull a weed, she is reminded of her childhood. In the spring she and her siblings would run barefoot through the fields. The watery sun would warm the soil, and early life would erupt from the ground. As the shoots became stronger the kids would rake off the protective straw so the little plants could drink in the sun. She knows that John and Jim are not that different from those tiny slips. They too need room to grow.

She met her husband during college. Their decision to marry was hastened when he was invited to Japan to study. Jack had been raised by his adoptive parents, the McGilvreys, a Scottish family. They took him in as a baby after answering an ad placed by his natural mother, who couldn't afford to raise him herself. Mary never met her adoptive father-in-law, who died while Jack was still in high school. Shortly before the wedding her father came to the young man with a suggestion. "McGilvrey is not a Japanese name.

Maybe you should change it to something more appropriate now that you are getting married." He didn't have to add *to my daughter.*

Jack wanted to please his new father-in-law. The *Mc* in Mc-Gilvrey sounded a bit like Maki, a common Japanese surname pronounced Mah-kee. "What do you think, Mary?"

And so they became Mary and John Maki. Officially he was John M. Maki. His middle name was now McGilvrey, a tribute to his adoptive parents.

Mary opens the front door and blinks through the blinding snow. Children in bright winter coats fly down the steep sidewalk. The white scene looks like an Impressionist painting. Shouts and laughter ring through the moist, cold air. The boys are fine. She goes back in the kitchen to put up her soup.

Winter 1957. New York City.
Night has fallen on the East Coast. Eight-year-old Joseph Helfgot is eating dinner at a small table in the back room of a tiny corner grocery store on Manhattan's Lower East Side. The room is unheated and freezing cold so nothing can spoil the merchandise, says his mother. In the summer she stops selling cold cuts because it's too expensive to keep the refrigerator going.

The boy pushes his fork around a plate of matzoh brei, a horrible dish his mother makes at least every other day by shredding stale Streit's matzoh into a heavily oiled pan on a small electric burner next to the table.

The only time Rachel Helfgot cooks at their apartment is on Saturday morning, when the store is closed, and she hopes God will forgive her for lighting the stove on Shabbos. On Friday nights Joseph's father hurries off to shul down Avenue B for the evening prayers, taking the boy with him. Rachel stays at the store with her oldest child, Pauline, where they light the Sabbath candles and say the Hebrew blessings in the back room.

There's a Catholic church around the corner, and this once-Jewish neighborhood is now home to Italians, Poles, and a growing number of Puerto Ricans. The store is busy on Friday nights because workers get paid on Fridays, and although it's the Sabbath Rachel can't afford to close. Friday night is when the goyim buy, and her family needs the money.

Joseph picks through the matzoh brei, looking for a piece that isn't burned. The leftover matzoh spits and burns in the pan while Rachel runs out to the front of the store to wait on a customer. Although her husband is there, Rachel always adds up the order on the back of a paper bag, jotting down the numbers with a tiny pencil and adding them up at lightning speed. She can't read, but she learned how to add. She says Naftali, her husband, always makes mistakes. He lets customers buy on credit, but Rachel never does. Invariably she smells the matzoh brei burning, and she runs back to the pan, whips up an egg, and tosses it into the mess.

"Yosel, *zine* dinner. Eat a *bissel* more."

His friend Iggy's mother makes really *good* matzoh brei, sweet and light with cinnamon and sugar mixed in. Iggy's mother has her hairpiece done every other week. Rachel has hers done just once a month and lacquers her blond wig with large quantities of hairspray to hold her over. Iggy's family has their own bathroom, but the Helfgots have to share theirs with another tenant. It's cheaper.

At the table Joseph traces words into Hebrew letters in his blue notebook. He wants to finish his homework tonight so his father will let him play stickball tomorrow. After school today, before it got dark, he and his friends cleaned up a spot at the bottom of the park and tomorrow they'll play. Joseph writes the letters, but his fingers are cold, even with gloves on. *Alef, bet, gimmel* . . .

There's a litter box under the table and the room reeks of cat urine. It's nine o'clock and he longs to go back to the apartment, but his mother has decided it is safer for him at the store. She has set up a cot in the room that Joseph used to share with his sister. Now Joseph shares the room with a man who is thin and smells bad. Pauline's bed is in the parlor. The thin man smokes and snores, and his breath smells funny, like Naftali's when he drinks wine. Joseph is afraid of him, but the boarder pays Rachel five dollars a week to sleep there. After school Pauline stays upstairs at a neighbor's until her parents come home from the store. Rachel doesn't want her kids to be alone in the apartment with a strange man.

At night, while the thin man snores, Joseph tries to dream up ways he can make money so his mother won't need to have a stranger in the house. But he is too young to work.

His father pushes the curtain aside and enters the cold room. "You study?"

Joseph holds up his notebook. "See? I'm almost finished."

"*Gut zin.*" Good son. "I go to du bank now." He jingles a canvas pouch holding the night receipts. He leaves from the back door, going through the alley rather than calling attention to his departure out on the street. He heads for the night depository up on Avenue C.

The moment he's gone Joseph jumps up and races through the curtain. Enough studying. He has been sitting for three hours. The front of the store is deliciously warm and smells like ground coffee. Rachel is putting things away for the night. Less than five feet tall, she stands on a stepladder to make room on the shelf. Joseph starts to munch on leftover pieces of broken cookies that she always puts out on a plate for the customers.

"Yosel, give me dus cans." He hands her tuna fish tins from a display next to the cash register. Rachel reaches high up into a cupboard, the number tattooed on her right forearm at Auschwitz visible as she shoves the cans to the back of the shelf.

Two young men have come in. They begin to collect items from around the store, expensive things like toothpaste and soap. They pile them on the counter and then go back for more. It's a large order, more than thirty dollars, Joseph is thinking. He is happy for his mother. Nobody ever buys this much.

Rachel has been watching from her ladder and thinking the same thing. No one in this neighborhood has that kind of money. Three bottles of shampoo? "Another can, Yosel." He knows from her voice that something isn't right.

"Done!" she says too brightly, coming down off the ladder. She begins to add up the order, writing on the back of a paper bag, placing each item neatly inside another bag as she goes along.

"Yosel, put dus into the other bag," she says, reaching down under the counter. His eyes widen as she slips a butcher knife under her apron, gripping the handle tightly through the cloth with her right hand.

One of the men is perusing the coffee selection, picking up cans and setting them down again while his friend goes outside. Through the window Rachel sees the man look up and down the street while he lights a cigarette. His friend continues to study the back of a coffee can. The man outside looks up and down the street as he takes a drag. She moves quickly from behind the counter to the door, where she snaps the bolt shut, locking it.

"Put dus can back on the shelf or I kill you," she says in a low voice. She draws the butcher knife from under her apron and takes a step toward the customer.

Joseph's hands are sweating as he clings to the edge of the counter. His mouth is dry. He can only partially see what is happening, just his mother standing there with the knife. Shelves block his view of the man.

"Lady, let me out!" Joseph hears him say as something heavy hits the floor. Joseph watches a coffee can roll into the center of the aisle.

Rachel unlocks the door and pushes it open. "I see you again, I kill you," she says as he slips past her.

The two men run off down the street.

"Yosel, pick up dus can."

chapter three

Monday, April 6, 2009, afternoon. Intensive Care Unit,
Brigham and Women's Hospital.

an ICU nurse calls his off-duty colleague. He wants to be the one to tell her. In September they had gone together to Jacob Helfgot's bar mitzvah. Had there been enough room in the temple, Joseph would have invited the entire hospital.

"Lisa, it's Kevin." He hears a car honk. "Are you driving?"

"I'm on the expressway."

"I have to tell you something, some bad news. Don't wreck the car."

"What?"

"Joe won't wake up."

She is silent while she tries to absorb this.

"Who's on?" she finally asks.

"Jim's in there, talking to Susan. The organ bank is here too."

"What happened?"

"He threw a clot. Maybe a few of them."

"I'm on tomorrow."

"You don't have to take him, Lisa."
"Yes, I do."

Why is this room so dark? They've spent millions on floor-to-ceiling glass to keep us from feeling claustrophobic and they've got the damned blinds closed. I distinctly remember opening them. Then somebody dragged me downstairs for coffee. I didn't want coffee. I pull on the beaded chains and light streams into the room, spilling over Joseph's bed. The nurse looks up from his chart.

"Do the blinds have to be closed?"

"No, Mrs. Helfgot. It's just respect for your husband."

If it's dark right now, I will die. And I have never been able to get this nurse to call me Susan. He's always so shy, and today especially.

More people are coming into the room, some of the cardiac doctors. One says, "I'm so sorry, Susan."

"I know. It's okay."

"Your husband was a wonderful man"—*wait, I've already heard it, please don't say it again*—"funny, brilliant, noncompliant, intense, original, outrageous"—*or another phrase*—"a boost to morale, one of a kind, in love with you, passionate about his kids"—*or something. I'm not listening.*

It's not that these words are inaccurate. It's that nobody has spoken the deeper, more important truth: "Your husband was so alive. And now he is so dead."

Jonathan, my stepson, is sitting on the long couch along the far wall. He took the red-eye from Los Angeles on Saturday night to be with his dad during the transplant. He slept late this morning because of the time change and arrived at the hospital expecting to find his father awake. I didn't call him. I waited until he walked into the room to tell him. I wonder if he's angry that I didn't call and make him race over here. But for what?

Now we're waiting for Ben to arrive from school. He'll graduate in a few weeks and start New York University in the fall. Joseph was over the moon about that. When I walked into the kitchen last December,

Joseph was sitting at the table with his metal heart pump clacking away a mile a minute—whoosh, click, whoosh, click. It could speed up or slow down just like a real heart. From the sound alone I knew he was excited about something.

"I got in! I got in!" Ben was jumping up and down like a little boy.

Joseph was crying, but then, Joseph cried at everything—including, I swear, a Charmin toilet paper commercial.

"How's Dad doing?" Ben has called his mother from school.

"Come over to the hospital when you get out of class, okay? Jon's here. There's a lot going on."

"Okay. I'll be there around three." Ben isn't worried. There was always a lot going on with his father.

"Love you, sweetie. Be careful."

As Jon and I wait for Ben to get here, I shake hands and hug people and thank them for stopping by. Dr. Lynne Stevenson, Joseph's cardiologist, walks in, regal and aloof. Her eyes are swollen and red. Joseph was angry that she didn't come to Jacob's bar mitzvah, but he got over it.

"You really won't be there?"

"No, Joseph. I don't mix with patients outside the hospital. I just don't."

"You really mean that? After all this, everything that's happened, you still won't come?" She has saved his life more than once.

"No, I'm sorry. I can't."

Lately, though, they've grown a little closer. She is holding something in her hands, and I recognize the videotape of Joseph's mother describing her life in Auschwitz, although I'm not sure "life" is the right word. Joseph lent it to her when he learned that her daughter was writing a paper for school on the Holocaust. She tells me they watched it together. She hands it back to me, tying up loose ends because she knows I won't be coming back. Experience has taught her to see through this moment and out the other side. It's true, I guess. I won't be coming back. I never imagined that I'd miss this place, but as long as I had a reason to be here, it meant Joseph was alive.

She hands me a card and says, "Don't open it now."

The room is crowded and getting too loud. Jon is on the phone, trying to reach his older sister, but Emily is in Paris on vacation, where her cell phone doesn't seem to be working. Now he's talking to Joan, his mother, informing her that her ex-husband is dead. I wish I had thought to call her so she could brace herself to comfort her children. I'll call later.

Joseph is on the bed, his chest falling and rising, down and up, the respirator pushing air in and out until Emily gets here and we work everything out. Maybe we can donate his kidneys. The doctors seem to think so. They told me someone would be here to talk about it, probably tomorrow.

Suddenly Ben is standing in front of me. He sees all the people in the room, and he knows.

"Are they sure, Mom?"

"I'm so sorry, Starshine."

He tries to be brave, but his chin and chest are quivering. Of all the Helfgot children, Ben is the one with the coolest demeanor, at least externally. It must be the one-quarter British phlegm from my side of the family.

The doctors have been waiting for Ben. I'm not sure why, but they are leading us into a conference room.

Susan and the two boys pass within a foot of Esther Charves as they are led down a long hallway. She tries not to meet the family until the doctors have explained that there is nothing left to be done for their loved one, still pink and warm to the touch, but dead nonetheless. Sometimes families cannot grasp it. They ask for one more test, one more procedure, anything that might alter or at least defer this terrible truth. Esther will sit down with Mrs. Helfgot only after the meeting where any lingering questions the family has will be answered. Meanwhile she will continue to gather information, trying to learn everything she can about the potential donor. Sometimes that's impossible. Sometimes a driver takes a slippery curve on a dangerous road and dies in a strange hospital, far from home.

This case is very different. She knew it would be the minute she arrived and saw the expressions on the faces of the doctors and nurses. Esther has learned the basics—Helfgot's line of work, the names of his children, and so on—but never has she put together such an intimate portrait of a potential donor in just a few hours. She has been collecting comments all day about a man who must have been easy to know, who had insisted on being known.

"Mr. Helfgot drank only fresh-squeezed juice. His wife brought it every day. He loved fresh juice of all kinds, and she wanted to make him happy."

"He was an intellectual who also loved Judge Judy and Costco. He called it applied sociology."

"One day he had Chinese food delivered to the whole floor. Sometimes pizza on Friday, or big baskets of candy or cookies."

One of the nurses told Esther about the day Helfgot noticed that she was feeling down and asked her if something was wrong. She confided that her fiancé had broken off their engagement. "What, is he gay? He must be gay. In your case there's no other explanation. Don't worry. There's someone else around the corner."

Esther's colleague Meredith Pitzi has been plowing through Helfgot's medical records. She wonders whether the years of heart-failure medication may have compromised his kidneys and liver, making them unsuitable for donation. This will be one of those times when they won't know for sure until they inspect the organs at the time of recovery. The organ bank has found a potential recipient in another region who is in desperate need of a heart and who is a match. Couper is on the phone with his doctors now. When the call is over there's a quick huddle. "They want the heart," he says. "They want to know when."

Good question. Meredith and Esther know it's too early to talk about *when*. Wheels are already in motion in case this is the face they've been waiting for, but they don't want to mention it yet, not even to the doctors. They want to be certain.

"We'll know more in a few hours," Esther tells Dr. Couper. "You're sure about the heart? I don't want to say anything to his wife until you're sure."

Families dealing with sudden death often find solace in organ donation, knowing that part of their loved one will live on. Esther doesn't want to raise this hope and then have it shattered. This family has been through enough.

Because Helfgot is a registered organ donor, Esther doesn't need his wife's permission for his kidneys or liver, or even the new heart. But a face is something else entirely. She will ask Mrs. Helfgot only when she is pretty sure the answer will be yes. Slow down, Esther, she tells herself. One step at a time. Even Dr. Pomahac, the surgeon waiting to perform the nation's second face transplant, doesn't yet know that they may have a face.

All day long she has watched Mrs. Helfgot through the glass, greeting visitors. They come in, approach the bed, and touch Joseph's hand or his shoulder. Or they grasp the bedrail and gaze at him in silence. Some of them have broken down, but not Mrs. Helfgot. She seems strong. Given what may soon be asked of her, she had better be.

Why are we in a crowded conference room? I want to go back and sit with Joseph. Who is that bald man in the blue shirt with his sleeves pushed up? He looks about my age, but in this pressured place, I wonder. A week in the tropics might take ten years off his face. He's a neurosurgeon, and he's going on about the swelling of the brain and how they could go in there and relieve the pressure, but it may not . . .

"Excuse me," I butt in. "Go into his head? Why would you do that?"

He clears his throat. "Well, as I said, to try to relieve the pressure."

"But he's dead," I hear myself say. Ben starts to cry.

"We just want you to feel comfortable. If you would like us to try something . . ."

The room becomes unbearably silent. No one wants to be the first to make a noise, even to scrape back a chair and stand up. That would mean moving on, and no one here is up to it. There has always been something else to try, but that's not true anymore.

When the room clears out I tell Jonathan and Ben that we have to go home and tell Jacob, the youngest Helfgot. Ben will ride with Jon. They need to be together.

"If you guys get home before me, don't say anything until I get there."

I work my way through the hospital complex to reach my car. Each building is an achievement, a dramatic statement underscoring the power of the institutions and contributors that have created this huge hospital: the Shapiro Center, the Tower, the old Women's Lying In, the Amory, the bridges connecting Children's Hospital, Harvard Medical School, and Dana-Farber Cancer Center. There's a billion dollars of real estate here and the best technology money can buy. But none of it could save my husband.

The walk soothes me. Finally I am alone and anonymous. All day people have been staring at me. How is she holding up? I know they mean well. I'm on the Pike now, the hospital's long pedestrian spine that connects all the buildings. Connors Health Center for Women? Take Exit 4. Amory Building? You can go up the elevators this way, ma'am, or cut through and go that way.

Right now I'm going that way, toward home, to tell Jacob. I don't know what I'll say, exactly, but it will probably be sealed in his memory for the rest of his life.

"Mrs. Helfgot," someone calls as I clear Exit 7. It's one of the thoracic surgeons. "I heard Mr. Helfgot got his heart!"

I guess the news hasn't made it to Exit 7. I paste on a smile and walk swiftly past him.

At the end of the Pike I enter the original Peter Bent Brigham Hospital. It smells of old wood, its stately Roman columns visible through the revolving door. Under a dimly lit alcove sits a case holding the 1990 Nobel Prize in Medicine, which was awarded to Dr. Joseph Murray, who performed the world's first successful organ transplant right here, at this hospital, in 1954. It was a kidney, and he transplanted it not in an operating theater, but in a quiet room away from probing eyes that would surely widen in horror at the idea of taking a warm organ out of one body and putting it into another. But what is unthinkable in one decade can be commonplace in another.

I have come here often when it was all just too much. There are pictures and display cases holding old instruments, and it brings me peace to be reminded that Joseph and I are part of a much bigger picture, that the struggle to keep him alive, so enormous to us, is a soft breeze blowing a tiny ripple across a vast ocean. I glance quickly at the alcove as I head out the door.

Luis, the valet, hands me my keys. He has been so good to me, standing in snow and rain and blistering sun through the seasons of my years at the Brigham.

"Buenas noches, señora. ¿Señora?"

"¿Sí?"

"¿Estás bien?"

"No, no estoy bien. Perdona."

To this day I don't remember driving home, or what I said to Jacob when I got there. He sat on Jon's big lap and cried while his brother held him for a long time until both their tears were spent. Jon told me I was brilliant, and for this I will always love him. He now admits that he can't recall a single word I said, just that it sounded right.

chapter four

Tuesday, April 7, 2009, morning. Intensive Care Unit.

Lisa Kelley, Joseph Helfgot's intensive care nurse, stands over her patient, wiping down his arms with a cool moist towel before moving on to his legs. She's keeping him clean as she helps him through this one last ordeal. Last year on a Tuesday morning, they would have been playing cards. They played a lot of Crazy Eights after a bad reaction to heparin required the amputation of Helfgot's toes; card games helped him exercise his mind after an unusually long postoperative delirium. It took him a week to tell a diamond from a heart, and another week to remember that eights were wild. After that Lisa stopped letting him win.

She was here two days ago, on Sunday morning, as the Helfgots waited for the transplant operation to start. Like many of the doctors and nurses who knew the couple, she had popped in to help them pass the time during the interminable delay. It wasn't a party exactly, but the mood was anticipatory, relaxed, and happy. Given what Joseph Helfgot had endured, and how strong he looked, no one imagined he might not make it through the transplant.

Was it really only two days ago?

"Lisa, lovely Lisa," Helfgot sang. "What brings you here?"

"I work here, remember? So, are you nervous?"

"I am, but excited too. I feel like I'm going down a tunnel. I'll be really happy when it's all over."

Her eyes well up as she recalls their final conversation.

"Lisa? Do you know when Mrs. Helfgot is planning to get here?"

Esther Charves has spent the night in the hospital. She had hoped to grab a few hours of sleep on a couch, but she and Meredith were up all night, drinking coffee and consulting with doctors on the organ bank's list about the potential suitability of Helfgot's organs. Organ bank policy is to wait up to an hour for each doctor to return the call before they move on to the next name. It makes for slow going.

"If you're ready to talk, I'll call her."

"I'm ready." Esther will start by asking about the heart. If that goes well, she will ask Mrs. Helfgot how she feels about donating her husband's skin, which can be used in reconstructive surgery. And if *that* goes well, she will move on to the big question—but not right away. Her immediate priority is meeting Susan Helfgot and supporting her, no matter what she decides. For now, Esther will put the face transplant out of her mind.

Dr. Greg Couper spots Susan entering the ICU. They haven't spoken since right after the heart transplant on Sunday night, when he plopped down in a vinyl easy chair in the visitors' lounge, his hands dangling tired and free over the armrests, and told her the surgery had gone well, but there had been a clot in the aorta. Her stepson, Jon, was with her, and Susan hadn't asked about the clot. Then Couper went off to start a second heart transplant. She couldn't imagine how he was able to do two of these in succession, but there was no choice. Life and death don't happen on a schedule.

Couper had first met the Helfgots eighteen months earlier, when he installed a VAD, a ventricular assist device, in his patient's chest. Joseph came into the hospital extremely ill and was put at the top of the region's heart transplant list. But when a heart never came, and Helfgot's own heart finally gave out, Couper implanted the VAD to buy them some additional time. It turned out to be all the time they had.

For the first few months after VAD surgery Joseph dropped off the heart waiting list while he regained his strength. When he was healthy enough for a transplant operation, he went back on the list.

A simple machine, really, considering its value, a VAD diverts blood out of the heart and into a metal container about the size of a canteen. When the container is full, a metal plate inside is pushed down by a motor, forcing blood through a tube and directly into the aorta, the main blood highway. From there the patient's blood branches out to the rest of the body.

VADs are lifesavers, but they usually don't last more than eighteen to twenty-four months. And they come with major restrictions. Patients must stay away from electrical outlets, and even blankets or clothing that could cause static electricity and short out the wiring. Forget about taking a shower or a bath.

If the machine malfunctions, an alarm screams out and often the patient must be hand-pumped manually, usually by a family member with a cool head, until they can get to the hospital. Helfgot's VAD had malfunctioned twice. Susan will never forget her frantic middle-of-the-night phone call to Dr. Couper when the alarm sounded. The diagnostic testing she tried showed nothing was wrong, but something clearly was. Couper talked her through it, and the problem turned out to be a faulty electrical cable that needed repair. The boys had awakened from the alarm, and rather than scaring them further with yet another sudden dash to the hospital, they switched to the external battery pack and waited for morning to go into the hospital to have it fixed.

For the next three nights Susan couldn't sleep. Even when everything is working perfectly, VAD life is a special kind of hell. But it's better than dying.

This morning Susan greets Dr. Couper and waits for him to say he is sorry. Everyone has been repeating this morbid little mantra, and she has come to expect it.

Instead he rubs his white goatee and says, "I gave my action figure to my little nephew."

Susan has forgotten all about that. Around the time Couper installed the VAD, Joseph was working on the marketing of *Iron Man*, starring Robert Downey Jr. as a brilliant scientist who almost dies in an explosion. To survive he builds himself an iron heart. Joseph used to call Dr. Couper the real Iron Man because he put metal VADs in people.

Just after the movie came out, a deliveryman showed up at the ICU reception desk with a large box on a dolly. It sat there for a while until the charge nurse got tired of tripping over it and finally ripped open the top with surgical scissors. A typed note lay on top: "For Joseph Helfgot's hospital friends, from Paramount Pictures." The box was filled with T-shirts, action figures, coffee mugs, visors, and mouse pads, all bearing the *Iron Man* logo. The action figures went first. Some of the doctors fought over them like little boys.

Couper puts his hands on Susan's arms and squeezes tight— his way of giving her a hug. He is too formal for anything more. He finally says the words she has been expecting and even waiting for.

"I'm so sorry. The clots. He was probably gone eight minutes in. We just didn't know."

Dr. Rawn has joined them. She looks at them both and says, "Please, don't feel bad. It's nobody's fault."

She wonders what it must feel like to work so hard and fail. They must go through this type of frustration fairly often.

"Mrs. Helfgot?" Susan has been looking out the window at the sheets of rain running down the glass. She turns around and sees a short, middle-aged woman in a business suit standing at the door.

"I'm Esther Charves, with the New England Organ Bank." As Susan leads her over to see Joseph, she says, "I feel I know your husband. So many people have told me stories about him."

They stand by the bed watching his chest rise and fall. Susan begins telling Esther about her husband.

As they make their way to a conference room where they can talk in private, Esther learns that Susan and Joseph were together for almost thirty years.

"Your husband was a registered organ donor?" Esther begins. She is still waiting to receive a copy of his organ donation card from the organ bank.

"Oh, yes, he was really into it, especially after waiting so long for a heart. A few weeks ago a documentary film crew came to our house and interviewed us. Joseph was so happy that he could talk about organ donation."

Esther knows all about the film crew. The organ bank's in-house lawyer has been drafting language for a special consent document that Susan will need to sign if the face transplant happens. Face transplants are big news, and protecting the identities of both the donor and the recipient will not be easy. A television production team for a series called *Boston Med* has been at the Brigham since January. They have been filming patients at home who are waiting for organs, including Helfgot, and hoping to follow them through their transplant surgeries, if and when they occur. But the day Helfgot got his new heart they were off filming another transplant.

The producer has the hospital's permission to film the country's second face transplant, if and when it occurs at the Brigham.

With three weeks left in their five-month shoot, it looks like he may get lucky. But nobody could have predicted that the face would belong to Joseph Helfgot, a man they have already interviewed.

For now Esther tells Susan the most urgent news, that a potential recipient in another city is desperate for the transplanted heart.

"It can be used again?" Susan says. "That's incredible." She feels no connection with this newly transplanted heart. It is not her husband's heart, the one that beat in his chest throughout their days and nights together, the one she could feel and almost hear during their most intimate moments. This new heart never really became a part of him. But she knows that the family who donated it will feel a loss if they learn the recipient has died. The news that their loved one's heart might save someone else will be huge for them. Had it been Joseph's heart, she would have wanted to know that the gift had not been in vain.

"It's amazing, but apparently it can go to someone else. We're not sure about the kidneys and the liver. We'll know more later. Your husband was sick for such a long time."

"Whatever you can use, please take it."

"Mrs. Helfgot . . ."

"Please, call me Susan."

"Susan, sometimes we are able to recover tissue. How do you feel about that?"

"What kind of tissue?"

"Well, muscle, bone, veins, and sometimes even heart valves. Sometimes tissue is used right away, but it can also be preserved. The tissue on your arm, for example"—she holds out her forearm and runs a manicured fingernail along the inside. "And the back and the buttocks are often used for breast reconstruction after a mastectomy."

"Really? Well, whatever you need." Susan laughs. "I'm sure Joseph would enjoy the idea of winding up on a woman's breast."

Esther smiles. She knows it's too soon to ask for the face.

They've only just met. But she already feels that the answer will be yes.

"People really loved your husband."

Susan nods.

"One of the doctors told me he never saw anyone fight so hard to live."

"Yes, it was in his blood. His mother survived Auschwitz. She was such a fighter."

"Did you get to meet her?"

"I did. She was the toughest person I ever knew. And after what she went through to survive, there was no way Joseph wasn't going to fight."

"I think he fought for you and the children."

Esther leaves the room to call Chris Curran at the organ bank. "I just talked to Mrs. Helfgot. She said yes to organ and tissue donation. She even made a joke about tissue. I'm going to ask her."

Chris Curran swallows, and his mind shifts into high gear. If Mrs. Helfgot says yes to Esther's next question, there will be a lot to do, and quickly. And there's a heart going out of the region. Tricky.

"Are you sure?"

"I'll be really sure before I ask."

A little later Esther and Meredith are in the nurses' lunchroom, catching a bite and planning their next step.

"He matches up," Meredith says. "His HLA, antigens, age, everything. But Dr. Pomahac still hasn't seen the face."

Esther knows that, but she worries that if Pomahac visits the unit, the whole hospital will be buzzing. He's been on television talking about face transplants, and everyone knows that if the Brigham does one, he'll be the lead surgeon. There is no reason for a plastic surgeon to be on a cardiac floor. What if someone sees him and mentions his visit to Susan before Esther has asked her? What if somebody calls a reporter?

A few nurses enter the lunchroom, talking about a conference they attended yesterday. Helfgot was supposed to have been the guest speaker. The topic was "life on a VAD."

"They announced it from the podium," one of them says, "that Joe finally got his heart. But not what happened after that."

"Maybe there's something about the conference on the hospital's website," Esther says after the women leave. "It might be posted. If there's a picture, Pomahac can look at it."

A quick search shows that Helfgot's picture pops up in several places. They find a nice profile shot of him. Perfect.

A woman with creamy white skin and auburn hair walks into Joseph Helfgot's room. Pam Levine is his surrogate daughter. Her father, Sol Levine, hired Joseph to teach sociology at Boston University in the 1970s. Pam was a little girl when she first met Joseph. She was sitting in pajamas on her father's knee during a poker game in her parents' dining room.

"Daddy," she whispered, staring at Joseph's enormous Jew-fro. "Why does that man have so much hair?"

Years later Joseph hired Pam fresh out of college to work for his fledgling movie research company. She was a gifted marketer, but her real specialty was managing the boss. One of the major studios eventually hired her away, but she and Helfgot stayed close, becoming even closer when her father died suddenly in 1996. Although Pam is a glamorous Hollywood businesswoman, today, in a baggy sweater and jeans, she could almost pass for a graduate student. She has obviously been crying.

She and Susan embrace. "When I landed, I went to the house."

"Have they found Emily yet?"

"She's on her way." Pam looks down at Joseph and rubs his shoulder.

"He looks so good, Sue. How can he look so good?"

"I know. It's weird, isn't it?"

"Have you thought about what you're going to do?"

"We're just waiting for everybody to show up. The rabbi is coming tomorrow to help us say goodbye."

"I mean with Joseph gone. After."

"Oh."

What am I going to do? Finally finish my teaching degree? Sell the house? Take a long trip somewhere? God, I have no idea. I can start by tossing out all the Splenda and the salt-free sauces. But what am I going to do?

Pam's question terrifies me. I close my eyes to hide and see a color-less wall. It's so very, very close. Too close. I take a step back to see it better. Why am I holding my breath? I suck in air and the wall is gone. Everything is white.

5 p.m. Helfgot's bedside.

The widow has left to attend to matters at home. A neurological exam required to determine brain death was performed earlier in the day, and the time has come to pronounce Helfgot officially dead. An intimate group gathers. Ideally such a group includes someone who is unknown to the family. Today Dr. Gentian Kristo, a surgical resident, will follow to the letter the protocol established by the hospital in keeping with the Uniform Determination of Brain Death Act. Dr. Couper will sign off as the attending physician.

This is all much more complicated than it used to be. Advances in modern science have made it possible to keep a body alive after brain death: ventilation machines deliver rich air; IV lines pump fluids and medicines keep hearts beating and maintain blood pressure; feeding tubes through the stomach deliver sustaining nutrients. Even when the mind is gone machines controlled by other minds can keep a body stable, suspended as if in a cocoon, ghostlike, hovering between this world and the next in a permanent vegetative state.

The heart keeps beating after brain death. Unlike other nerves in the body that are attached to the brain stem, it has its own rhythmic nerve center and can keep going without instructions from the brain. All it needs for its cells to keep working is oxygen. As long as a machine can supply the lungs with oxygen, a body can be kept "alive" for some time. In the famous (but unusual) 2005 case, Terri Schiavo's body was kept "alive" for fifteen years until a court demanded that the ventilator be turned off.

A strong light is placed within millimeters of Joseph's eyeball. His large dilated black pupil, ringed by a hint of blue, remains fixed as he appears to stare straight ahead. Pupillary reflex—gone. A doctor turns Joseph's head and checks for doll's reflex. But his eyes do not move.

Next Dr. Kristo lightly touches a cotton swab to Helfgot's closed eyelid. There is no twitch, nothing. Corneal reflex—gone.

Then he lightly stabs the skin in sensitive spots with the sharp end of the cotton swab. No response. They continue checking reflexes until they come to the final test for apnea. The ventilation equipment is turned off. The room becomes silent. A minute goes by, then another. Nothing. No gasp for air, no primordial struggle for survival. Joseph's lungs are screaming for oxygen, but the cries go unheard. His brain is gone and no longer commands his body.

They quickly reconnect the ventilator because there are still goodbyes to be said and lifesaving organs to protect for others. For a little longer, Joseph Helfgot will float between worlds.

At 5:13 p.m., on Tuesday, April 7, 2009, Dr. Gentian Kristo pronounces Joseph Helfgot legally dead. Cause of death: cerebrovascular accident. The two doctors sign the death certificate and go off to care for the living.

"Chris," says Esther, "it's me again. They've declared him. His wife went home for a while. She'll be back around seven. I told her we would do the oral history with Meredith when she returns. I'll ask her then."

"We should tell Dr. Pomahac."

"If you do, let him know it would be better not to come up here until I talk to the family."

"I'm all over it."

Dr. Bohdan Pomahac, the director of Brigham and Women's Burn Center, is on his way out the door. It's early for him, but his wife is going out with her friends, a rare event. With two small children, she barely gets a moment to herself. His phone starts to buzz. The words "I'm sorry, but whatever it is, I have to go home" are already half-formed on his lips. But when he sees the number on the screen, his mouth goes dry.

"Dr. Pomahac? It's Chris Curran. We may have the donor."

"Where?"

"Right in the hospital. He's a heart transplant that didn't work out. Esther Charves is there. The family has already said yes to tissue. Esther hasn't asked for the face yet. His wife is coming back around seven to do the history. We'll know then. I'll call you the minute we know for sure."

Pomahac hits a name on his speed dial list. Julian Pribaz, director of the Harvard Plastic Surgery Resident Program, was his teacher and is still his mentor. A year ago the two men flew to Brussels and trained together to perform this almost unprecedented operation.

"Julian," Pomahac tells him, "we may have a face."

chapter five

Tuesday, April 7, 2009, early evening.
Brigham and Women's Hospital.

esther needs to clear her head. She glides down the escalator and is swept along with the crowd of people spilling into the street. Daylight Savings Time has kept the twilight at bay, but in the misty drizzle it feels like night. A strong cold wind smelling of the sea rips down Francis Street, but she doesn't mind; after thirty-six straight hours indoors she finds the fresh air exhilarating. She walks toward Huntington Avenue, collecting her thoughts and preparing herself for the conversation she is about to have with Susan Helfgot. She has had to do many difficult things in her years at the organ bank, but this one is in a category all its own.

Nothing in her life has been easy. She grew up near the sea in Bristol, Rhode Island, a quiet town that juts out into the cold north edges of Narragansett Bay. Bristol isn't far from the mansions and yachts of Newport, but only if you're measuring in miles. When Esther was little the town was solid working class.

Her father was a construction worker, and she remembers a time when they didn't even have a telephone.

As a teenager she would babysit for the children of an older cousin in a house filled with books and magazines, so Esther became a big reader. She won a partial scholarship to a good private high school, but she barely had time for homework. Because her mother was very sick Esther took a part-time job and helped care for her siblings. She has a powerful memory from early childhood: she is waving to her mother, who stands at the window of a Boston hospital. At the time children were not allowed to visit, and to Esther the hospital looked a lot like a prison. It was the Brigham in an earlier incarnation.

Esther went on to Boston College, and later to Boston University. She married early and had three children, but stayed in school and became a nuclear medicine cardiac technologist, performing isotopic scans at a big heart clinic.

A warm and compassionate woman, she was friendly with some of the patients who came in for imaging, often with their children. Some were waiting for new hearts, and although they knew the odds were against them, they remained hopeful. Over time a number of her patients died waiting.

It was a sad progression, and Esther grew tired of watching it. Eventually these patients were unable to walk, and they started arriving at the clinic in wheelchairs. A few months later they might be on oxygen. Finally there would be a missed appointment that was never rescheduled. Sometimes word would trickle into the clinic that one of their patients was given a new heart. Far more often, though, the story ended badly.

One patient, Jerry, was forty-seven, with five sons. Esther had grown attached to the boys, and when their father died in 1999 she attended his funeral. Jerry's five sons stood next to the casket, lined up as progressively shorter versions of their dad, right down to the little five-year-old, who stood there bravely, shaking hands. She heard later that the family was evicted from their home and living in a tiny apartment.

For Esther, Jerry's death was the last straw. By now her marriage was over, and her youngest child was already in college. She knew all too well what it was like to wait for a heart. Now she wanted to work on the other side of the equation, the supply side. The New England Organ Bank had an opening, and Esther was hired.

As a family services coordinator she sits with grieving mothers, weeping wives, and devastated husbands on the worst day of their lives. And she asks them for permission to open up their loved ones and remove organs that will help other patients, complete strangers, many of whom will die without these extraordinary measures. It is a very demanding job. Often, at such a moment, the person she's speaking to would say no to anything, even a gift of a million dollars, because her husband or his wife went out this morning to pick up a carton of milk and never came home. Sometimes all the survivors need is a little time to absorb the news that their loved one is really gone and to consider the request. She can offer them at least that.

She still remembers the first request she ever made. She was trembling. This was the job she wanted, but until then she hadn't fully understood how bold she would have to be. Like anything else, you get used to it—not completely, but mostly. Now she never flinches. She believes a family has the right to hear the question she is asking on behalf of the person who has just died. "I am gone," she imagines them saying. "Please let someone else have a chance to live. There's been enough tragedy for one day."

Esther is not a saleswoman. She embraces their noes and yeses with equanimity. She knows that some people want to have all the available information and will still say no, and she feels she owes them the same respect and dignity she would give to someone who says yes. She hopes families will say yes because lives are at stake, but also because she knows from experience that most families find solace in donating. The decision to help another family—complete strangers—may be the one thing they can hang on to in the brutal days ahead.

These conversations don't always go well. Sometimes family members don't support the wishes of the deceased who has signed an organ donor card. Sometimes they become angry at Esther. Back when these requests were made by phone, one woman shouted, "Don't ever call me again!" As if she would ever call again.

Once, a mother whose child had just died spat out, "Go away. Leave her be. If I can't have her, neither can you!"

Esther was stunned, because a nurse had told her that the eighteen-year-old had been obsessed with organ donation. She had raised money for the organ bank and hoped to become a transplant surgeon. The girl died waiting for a liver that never came.

Esther quietly gave the woman her card and left. She always gives her card, even when the answer is a defiant no. Maybe in the dead of night they will call with second thoughts. It's rare, but it happens. Six months later the mother called her. "You were right," she told Esther, crying. "Now I have nothing. And neither does anyone else. I'm so sorry."

One night the sister of a young accident victim slammed her fist down on the table, shouting over and over, "Show me the money! Show me the money!" as her fist kept time with her words. Then her brother joined in. They continued pounding and chanting as Esther backed out of the room, blinking away tears and humiliated by their behavior. She could still hear them on the other side of the ICU door as she waited for the elevator.

But more often than not the families say yes. And with that yes somebody, somewhere will receive a precious and anonymous gift.

"What do you think?" a colleague asked Esther as they drove to the meeting with Dr. Pomahac at the organ bank.

"I'm not sure I could do it. I can't even think about how I would ask somebody for his wife's face. What about a child's face? No, I don't think so," Esther said.

But then she met Pomahac, a quiet young man with a receding hairline and sharp brown eyes. Esther reads people for a living. She knew he was nervous, this unassuming surgeon with his soft European accent. He seemed worried about their reactions when he asked for help in finding facial donors for his new program at Brigham and Women's Hospital.

She couldn't miss his anguished expression as he described how some of his patients were unable to swallow, to breathe, or to speak. They endured surgery after surgery, with only modest improvements. Some of them were unable to go out in public because of their injuries. Some were young men who lost parts of their head on the battlefields of Iraq or Afghanistan. Some were marred by cancerous tumors. He showed pictures to demonstrate how muscle and bone from a face donor can rebuild a life in a way nothing else can.

He explained that it wasn't someone's entire face he needed, just parts of a face. But everyone in the room that day knew that to a grieving family, this distinction would mean little. Dr. Pomahac knew it too.

After seeing the pain in his eyes as he described his wounded patients, Esther knew that, if she had to, she could ask the question. But she sorely hoped she would never have to.

The organ bank drew up a script that she and her colleagues could use. But they all knew that any script they tried to follow would soon be abandoned. This was uncharted territory. They would have to follow the conversation, wherever it went.

Susan Helfgot's quick consent to Esther's first question, and her composure, her smile, and her willingness to donate her husband's skin tissue—all these things make Esther believe that Susan will say yes. But how is she going to ask? How is she going to phrase the question?

She hears Dr. Pomahac's words and sees a picture in her mind of the man he has been trying to help, a man who is horribly disfigured from a fall. How it happened, she doesn't know. She knows only that he is married and has a child. And that he can hardly breathe, or swallow, or smile. He can't even step outside without ridicule.

Susan, someone is suffering horribly. Can you help him?

She utters the simple prayer she always recites before she meets with a donor family. She asks God to help give her the right words, and to allow her to be helpful to these grieving people, even if they say no to her request. She takes one last breath of fresh air and heads back inside the building.

At the door to Joseph Helfgot's room she says, "Susan? May I speak with you again? There's something else I need to ask."

chapter six

Tuesday, April 7, 2009, 7 p.m.
Intensive Care Unit.

Susan has returned to the hospital with Pam, who doesn't want her friend to have to meet with the organ bank lady by herself. The three women make their way to the conference room, where Susan and Esther spoke earlier. The darkness outside is almost complete, but no one moves to turn on the lights. Diffuse light from the hallway filters through the glass panels. Pam curls up in a chair while Susan sits on a couch, facing Esther.

Esther reminds herself that she is here to ask a question that Joseph cannot ask of his family. Given everything she has learned about him, she is certain that if he could, he would grant her request. This belief calms her as she begins, speaking softly: "Susan, there is something I have to ask you, something Joseph didn't sign up for. When he signed the organ donor card, it was for organs and tissue. What I'm going to ask you is a big deal."

With Esther scattering caveats and throwing up warning signs, Susan wonders what's coming. She can't imagine what it might be, but feeling a need to help Esther along, she offers a joke. "Don't tell me you want his Jewish nose?"

Esther's head snaps back a fraction of an inch. Then, resisting the temptation to respond to Susan's surprising guess, she sticks with the script in her mind. "If I told you there was a man who was horribly hurt in a fall, who can't eat normal food or talk on the phone, who sometimes has to breathe through a tracheostomy tube, would you want to help him?"

Well of course she would. For more than a month last year Joseph had both a trachea tube and a feeding tube. It was awful. He wasn't even allowed to have a sip of water. She can still hear him saying, "Please, just a piece of ice. My throat is so sore and dry."

"Esther, what exactly are you saying?"

"We have a man who needs a face."

Pam's arms fly up and she gasps.

"Oh my God," Susan says. "You really do want Joseph's nose."

For a moment nobody speaks.

"I think I have to do it," Susan finally says. "It wouldn't be right not to. But I have to call my children."

Pam thinks her friend is too exhausted to make such a big decision so quickly. "Sue? Are you sure?"

"If Emily and Jon agree, and Ben too."

She calls home. "But not Jacob." He's so young, she is thinking. Let's not burden him with this.

Jonathan answers the phone, and Susan explains that she and Pam are with the organ bank lady. "Yes, they still want the heart." She asks him to get Emily and Ben on the line. Jacob is playing video games with a friend.

When Emily and Ben have joined the call, Susan begins: "Guys, listen to me. There's a man in real trouble who needs help. He can't eat or talk or breathe very well. He fell very badly and lost his face. Yes, his whole face. I don't know how. That's not important, is it? They want to know if we will let them have your dad's face to help fix him."

There is no immediate response. "Guys? What are you thinking?"

Emily asks, "What do *you* want to do, Sue?"

"I want to do what Daddy would want us to do."

"He would want to do it," says Jon.

Susan listens as Emily explains it to Ben. "Do you guys want to think about it for a while?"

"Thinking about it won't change anything," Emily says. "It's what Dad would want to do. Jonathan, what do you think?"

"I think so too."

Sue hears Emily talking to Ben, who says, "Yeah. We should do it."

"Okay, then that's what I'm going to tell her. We'll be home soon. I love you guys."

"We're going to do it," she tells Esther.

Esther steps out of the room and calls Chris Curran. "She said yes."

"Oh my God. Esther, are you sure?"

"No."

"*What?*"

"Kidding."

"I gotta call everybody. Bye."

Curran starts punching in numbers. He calls the head of public relations for the organ bank. He calls the head of the organ bank, and somebody alerts the chief medical director. He tells the tissue service manager to start organizing her recovery team. Next comes the organ bank's lawyer, who has been sitting at her desk reviewing the consent documents. "Chris," she tells him, "you can't give Dr. Pomahac the go-ahead until Mrs. Helfgot signs the consent. And Meredith still has to take the social and medical history. Something could come up, or she might change her mind. Maybe she won't sign. I'm faxing it over to Esther right now."

Chris knows it's not a done deal, but he is dying to call Pomahac. He knows he's waiting at home and probably climbing the walls.

8 p.m. Suburban Boston.
Bo Pomahac isn't quite climbing the walls, but he is watching the clock and wondering when his phone is going to ring. Hanka, his wife, has left for her dinner with friends.

"Daddy?" His little girl is looking up at him. "Daddy, are you listening?"

"What, sweetheart?"

"Daddy, I need a newspaper for school. We need a picture cut out of a newspaper."

"Okay, honey."

"*Daddy?*"

"What?"

"We don't *have* a newspaper. That's what I'm trying to tell you."

"Okay. Let's look for something we can use. Get your scissors and we'll find some magazines, and then it's time for bed, okay?"

He keeps looking at his cell phone, the time passing slowly, as he reads his son and daughter a bedtime story. He wishes Hanka were home. He wishes his phone would ring and that the kids were already asleep. Please call, he keeps thinking. Please call.

"Daddy?"

"What?"

"Daddy, *read.*" The kids giggle.

"Sorry."

• • •

8:30 p.m. Intensive Care Unit.

Meredith has joined them in the conference room, where the air is charged with an energy that's impossible to measure. Esther and Meredith, Susan and Pam—four conspirators bound by the still secret knowledge that something big is unfolding here.

Meredith spends an hour taking Joseph's social and medical history from his wife. Susan likes her. This nurse from the organ bank is a little unconventional, with a funky tattoo and rings in her ears—quite a change from the staid crowd that normally roams the ICU. She delves deeply into Joseph's history with very personal questions. "You knew him for how long?"

Susan winces to hear her husband referred to in the past tense, but she knows she'll have to get used to it. "Almost thirty years."

After some questions about foreign travel, she asks, "Did your husband use drugs recreationally?"

Pam bursts out laughing.

Esther says, "Susan, you can only answer what you know about."

"That's what I'm afraid of!" They all laugh.

"Well, he taught undergraduate courses at Boston University. One class was Drugs in Society, and the other was Sex in Society."

Meredith grins widely. "For real?"

For real. Years ago, as part of his research, Joseph had shadowed a cocaine dealer, and Susan knows that her husband—with the very nose she has been joking about—had been known to sample the goods. He sometimes smoked pot, but Meredith and Esther are concerned about the kind of drugs that require needles, which Joseph never even considered using. They also know that to be eligible for his heart transplant, he was examined very carefully.

After the history is taken, Esther brings in the release for Susan to sign. She scribbles her signature. She is spent, completely exhausted. It is done.

As they leave the hospital everything seems muffled, as it does when you're leaving a very noisy sports event. After dropping off Pam, Susan cries on the way home. She tries not to, but she can't help it. She doesn't want to look like a wreck for the kids, but after holding the tears at bay for all this time, she can no longer suppress them.

Twenty-four years earlier. 1985. Boston University.
An energetic professor in his mid-thirties stands outside Morse Auditorium in Kenmore Square. Inside, the audience of four hundred undergraduates is stirred up. Today, Helfgot has told them, they will meet a special guest, a woman named Michelle. She used to be known as Michael, and in a few minutes she will be telling them about her sex-change operation at Johns Hopkins in Baltimore.

Earlier this morning, as they left the house together, Helfgot had asked Susan, his girlfriend, if she would come to his class today. "Michelle will be speaking," he said. "You know, the woman I had on the show last week?"

When he isn't teaching, Helfgot is on the radio as a local version of Dr. Ruth Westheimer, the hugely popular sex therapist during the 1980s. His Sunday night show on WHDH has a big audience. It breaks up the monotony for drivers returning home from Cape Cod at the end of a summer weekend, or from ski trips in Vermont and New Hampshire in the brittle winter cold. The professor is a bit of a local celebrity, and a political hot potato for both Boston University and the Catholic Archdiocese of Boston, who aren't exactly charter members of the Joseph Helfgot Fan Club.

"I don't know," Susan told him. "I've got a lot going on today at work."

"Please? We can have lunch after class, okay?"

• • •

"Michelle, you go ahead. Sit where you like, but somewhere in the middle. I'll introduce you in a few minutes. We'll see how everyone reacts when you stand up."

Susan arrives a moment later. "I can't believe you talked me into coming," she says. But that's not really true. He's very persuasive. They head inside, Joseph leading Susan down the center aisle toward the stage.

"Professor!" a student shouts. "I can't believe what they can do!"

"Christ, they think I'm *her*!" Susan whispers.

"Exactly!" He grins impishly as he jumps onstage and grabs the microphone. "And doesn't she look great?" He beams down at Susan in the front row, who shoots back an angry look just as someone whistles from the back of the room.

"There! Who whistled?" Helfgot whips around and begins pacing back and forth, brightly lit in a spotlight, the mike loose in his hand. Phil Donahue meets Mick Jagger, and both sides of him are delighted to be up there.

"That's perfect. That says it all! How to get a girl. Amazing. You don't even need language. Just whistle." He pauses and thinks for a moment. "Okay, so it's birdcalls. That's what it is, human birdcalls." The students are transfixed. "Do you guys remember the famous scene from *To Have and Have Not*?" Apparently not. "C'mon, you know, where Lauren Bacall tells Bogart to 'just whistle' if he needs her?"

There are nods of recognition, but not many. "So if you were a linguist, what would you say?" And off he goes on a rant about sounds and sex and beauty, rolling them into one big blur. Everyone is laughing and students are shouting back. Susan is trying to be invisible.

"Today," the professor announces, "we're gonna talk about how beauty is in the eye of the beholder. It's time to meet our special guest." He looks at Susan and gestures for her to come up. But Susan remains seated, as he knows she will, and Michelle stands up and starts walking toward the stage. There are some audible gasps.

"What is the lesson?" he shouts out. In a recent class he mentioned that men are sometimes attracted to transvestites, but rarely to women who used to be men.

"Don't you ever pull anything like that ever again!" Susan says when the class ends, punching Joseph on the arm. It's not a playful punch, and she's pleased when he recoils in pain.

"You're a jerk," adds Michelle.

April 7, 2009, 10 p.m. New England Organ Bank.
Chris Curran is on a late-night conference call with seven of his fellow employees as they try to cover every base: travel arrangements, legal considerations, the needs of the tissue recovery team, publicity control, family support, and much more. There is a plane to charter for the out-of-town cardiac team, who will have to get to and from the airport. It's not clear yet whether Helfgot's liver and kidneys can be salvaged, but the potential recipients for those organs are local, thank goodness.

The organ bank's hospital liaison will have to work with the Brigham to reschedule surgeries and soothe ruffled feathers. Pomahac and his team will need two operating rooms that are near each other for a minimum of twenty-four hours. This is a big request, as operating room space is hard to come by. Tying up two of them for so long with virtually no advance notice will be tough on everyone in surgical services.

"When do we think this thing will actually start?" the liaison asks.

"The family is having a bedside service tomorrow afternoon," says Esther. "Any time after that, I imagine."

After a few more kinks are worked out, they ring off, each with a list of tasks that will keep them working a while longer.

At 10:40 Chris Curran calls Pomahac. "We're set," he says, and Pomahac sighs loudly on the other end. "Esther will call in a few minutes. We'd rather you didn't go in tonight. Maybe in the morning, early, when nobody is around. Everybody knows you. Is that all right?"

Pomahac's mind is racing. He needs to call his team, get them together first thing tomorrow.

"Fine." He hears the car in the driveway. Hanka is home.

Twenty minutes later. Intensive Care Unit.
Esther stands over Joseph's bed. "We did good," she whispers. "We got it done." Although she had never met him when he was alive, Esther feels she has been speaking for him, that they have a kind of partnership.

Then she calls Pomahac. "I'm with the donor," she says. "Meredith says everything checks out beautifully."

"Thank you, that's great. How does he look to you?"

"He's an attractive Jewish man. Dark hair. He was supposed to speak at a nurses' meeting yesterday, and there's a good picture of him on the hospital website. A nice-looking man with blue eyes."

"Where's the picture?"

"Search his name. It'll come up." She spells it.

"I'll take a look," he says. This is it. He feels it. This is the one, the face Jim Maki has been waiting for.

Pomahac looks at his watch. It's late. He'll call Jim tomorrow. Let the man sleep. At least somebody will get some rest tonight.

chapter seven

a huge wave slams against the hull of the ship, stirring the boy from his deep sleep. "John," he whispers from the top bunk. "You awake?"

Another wave crashes as his older brother sleeps on. Jim hops down, instinctively landing with his feet wide apart for balance. He shimmies over to the porthole and stands on tiptoe, peering outside. As the ship pitches wildly toward the sea he is pushed against the curved steel wall. Dark foamy water rushes up to meet his gaze. He is thrilled by the thought that the ship's window might dip deeply enough to actually touch the ocean. But just when it seems that they are about to kiss the sea, the swirling water disappears and up they go again, butterflies rising in his belly as they ascend to the top of the next swell. Now he sees only gray sky through the glass. Then, suddenly, they lurch, his skinny knees turning to jelly as he braces himself on the aquatic roller coaster.

"John, wake up! It's my birthday again!"

"What time is it? Go back to bed."

But Jim is pulling up his pants as he holds the metal bunk for balance. "I'm going up."

"It's too rough. Dad will be mad at you."

"Not on my birthday!"

During the night the ship crossed the International Date Line, so it's Jim's eleventh birthday all over again. How lucky can you get? His mother promised him a second cake. Yesterday they gave him white with chocolate frosting; maybe today it will be lemon, his favorite.

He gets his sea legs running down the hallway and bursts through the metal door. Skipping up the mesh stairs, he skids out onto the deck, which is slick from salt spray. It is windy out here, and barely dawn. An endless sea of blue stretches out before him and meets a strip of crimson that fades to blackness. Any minute now it will turn into sunrise.

During the war this ship was converted into a floating hospital. Now it is tired and showing its age. It's not really a cruise ship, because it also carries freight. But to Jim it's impressive, a huge rocking mountain.

When they sailed to Tokyo almost a year ago, it took twenty-one days. That's when he learned how to walk on a ship. The boys' father was being sent to Japan to study something important, their mother explained, a legal document that had to do with the war. Jim heard that his father had a Fulbright, a great honor for professors. The youngster, who loved words, thought it was a good name for a man who was so full of knowledge and so clever.

Now they are heading home to Seattle. Butterflies fill his stomach again—not the kind he gets when something exciting is about to happen, but the ones that make him sick to his stomach. Hoping the feeling will go away, as it sometimes does, he makes his way to the cracked Ping-Pong table. He's dying to play, but John is still in bed. A man stands against the railing. Jim remembers seeing him before. The ship is not that big.

He picks out a paddle from the basket under the table. He never played before this trip, but now that's all he wants to do. Even with the boat listing back and forth, he can return pretty much anything that comes his way. John isn't as good, though, and he gets mad when Jim smashes the ball straight at him. Sometimes he misses, and they fight over who has to retrieve it.

"I *told* you not to hit it like that. I'm not going after it."

"Yeah, well, you're the one who missed it."

"Then I quit," John ends up saying, throwing down his paddle. It's a ploy. He knows Jim will go after the ball so they can keep playing.

"Mister, you want to play?"

"No thanks, son. Where's your brother?"

"He's still in bed."

"Maybe you should go back down. It's pretty choppy. Does your mother know you're up here?"

"Yeah." Embarrassed by the lie, he edges away from the man.

He isn't sure whether he is happy or sad that they are finally going home. Tokyo was a strange place, with strange smells and even stranger people. A lady came in every day to help his mother clean. Everyone lives like that in Tokyo, or so he thinks. He and John had expected that the boys in Tokyo would look just like them. But they didn't. Neither do the boys back in Seattle.

In Japan there was a big baseball diamond at the end of their street. Jim would run down there every day after school to see if there was a pickup game. At home he plays shortstop and third base, but in Japan he was often the pitcher. In Seattle he plays on a Little League team, and in the spring he intends to pitch.

In Japan they had a television and the boys watched all kinds of programs. They don't have one in their Seattle house because they live at the bottom of a steep hill and there's no reception.

Once their dad spoke back to the man on the television set in grammatically correct Japanese. The boys laughed, because next to the real thing, Jack Maki's accent sounded pretty bad.

"What's so funny?"

"Dad, you sound silly!" they said, erupting into fresh peals of laughter.

They went to the American School, where the classes were in English. But Jim picked up a lot of Japanese. He would talk to anybody, including the man at the noodle stand who took the grubby coins Jim picked up from the train station floor. Their lives in Japan were so different! Back home, they mostly had to stay in the yard. But in Tokyo they took three different trains to get to school, and their father thought that was fine. Japan, he always said, was a safe place to live. Even so, Jim and John weren't supposed to stop anywhere on their way to and from school, though they often did.

Their mother would nod her approval when Jim recited Japanese words and phrases. Mary spoke the language perfectly. She grew up speaking it on her parents' farm.

"John, I guess you and I have tin ears," Jack Maki would say as Jim and Mary exchanged a few words. John never did pick up much Japanese. He worked a lot harder in school too. Jim got straight A's in fourth grade in Tokyo. John got mostly Bs, but everyone knows that sixth grade is a lot harder.

"I can't find Jim." Mary is holding an ice bucket and trying to stay calm. "John says he went up on deck."

Her husband purses his lips in a frown. "I'll take a look."

"Jack, it's very rough."

"Don't worry, he hasn't fallen over. How's John?"

"Seasick."

Jack Maki walks toward the stairs. It's always something with that boy. Why can't he stay put and just do what he's told?

Jack is happy to be heading back to his job at the University of Washington. He is even happier to be returning to the ordinary luxuries of home. Tokyo in 1959 has still not completely recovered from the brutal bombings during the war that left much of the city in rubble.

He will soon be fifty, and for the first time in his life creature comforts are important. That certainly wasn't the case the first time he stepped on Japanese soil more than twenty years ago. He and Mary had married quickly so she could go with him on his teaching fellowship. They had met the year before, and he couldn't bear to leave her behind.

He can't quite believe how they managed in one tiny room, with a cold-water faucet in the corner and a gas ring heater that demanded a steady diet of coins to provide what little heat there was. He bought a small iron grill called a hibachi, and between that and the heater they managed to get their room up to about fifty degrees in the winter. The house had one bath and no hot water.

War broke out between China and Japan during their stay, but they weren't aware that the winds of a much greater war were gathering around them. When Jack's mentor in Seattle suffered what turned out to be a fatal heart attack, they rushed home and Jack took over some of his classes. From then on, his future at the university was assured. Years later Jack learned that intelligence officers from the Japanese Imperial Navy had been following him during his stay in their country to see if he might make a good spy.

When war broke out everywhere, Jack came to understand the world in a new way. Although he had been picked on as a boy for being Japanese, being adopted by a Scottish family had partly protected him from prejudice. But not always. Once a farmer told Jack's Boy Scout troop not to hike through his field. The scoutmaster argued with the man, who didn't want a "Jap boy" on his property. In the end the group retreated and had to walk all the way around. It was worse in college. When Jack signed up for ROTC, the Reserve Officers' Training Corps, he was rejected for being "not qualified."

He had planned to major in journalism until the dean cautioned him that no American newspaper would hire an ethnic Japanese. He switched to English literature, where he excelled. For his foreign language requirement he picked Japanese "for no reason other than idle curiosity," he wrote later. When the secretary of the English Department advised him that with his Japanese face he would never get an appointment teaching English literature, he switched to Japanese literature, and eventually to Japanese history.

After Pearl Harbor he and Mary were forced to pack up their Seattle apartment and move into army barracks in the town of Puyallup, squeezing into a tiny space with five other families. Seven thousand Japanese Americans were evacuated to Camp Harmony, with new arrivals being assigned to specific areas of the camp, depending on their education and background. "We weren't fearful or angry," Jack wrote later, "since it was accepted as a wartime necessity. Even though we were Americans, we were inescapably identified with the enemy, and we accepted it." He pointed out that it didn't even occur to them to demand fair treatment. This was well before the civil rights movement of the 1960s empowered many minority groups. It was a very different time, and the world was at war.

After a month at the camp, in the spring of 1942, Jack and Mary were sent to Washington, D.C., where they worked for the government, reading and analyzing enemy propaganda and Japanese radio broadcasts. All around them other analysts were doing the same for communications out of Germany and Italy.

After the war the State Department sent Jack to Japan, where he was stunned at the level of destruction. His job was to write a report on all the fractionalized government ministries. He helped monitor the first election, and he wrote to Mary every single day. Back home he earned his doctorate at Harvard and then returned to Seattle and settled into the life of an academician. He and Mary tried but were unable to have children. They adopted John, who was born in 1947, and then Jim, in 1949, when he was seven months old. Both boys were half-Japanese.

Up on deck Jack finds his younger son leaning against a wall.

"Jim, where have you been? Your mother is worried."

"Nowhere. I don't feel too good, Daddy."

"Go lie down," Jack tells him. He is angry, but also relieved. He too was a little concerned but was able to hide his apprehension behind his wife's.

Down below, Jim says, "John, do you think we can play Ping-Pong a little later?"

"Jim," his mother says, "please stop talking. You boys should lie still and try to rest."

"But it's my birthday!"

"Okay," says John. "We'll play later. But no smashes, okay?"

"Okay."

Wednesday, April 8, 2009, 9 a.m.
A veterans' home in central Massachusetts.
Before leaving his room Jim Maki pats down his dark brown, arrow-straight hair with his good hand. Although he's fifty-nine, there is no gray. I may have no face, he thinks, but my hair's still good. Down in the dining room he sits in one of the clunky captain's chairs, ready for the weekly house meeting. The residence is run by a private charity that helps Vietnam veterans with medical problems. These are men who have nowhere else to go. They are all disabled, and they have all struggled with drugs.

The room is covered with washable vinyl wallpaper with a loud striped pattern in white and Kelly green. A cheap print of a man with long hair in a brown robe hangs from the center of one wall. His hands are tied together with rope and he looks to be in pain as he gazes up to the heavens. All that's missing is the crown of thorns, or perhaps a halo. As the meeting begins, one of the residents is complaining that his room is too hot at night.

When it's Jim's turn to speak, he addresses a man across the table. "I gave you razors," he tries to say, "and other stuff too. But then you went and took my ——." It's hard to make out what he's saying. He tries to say each word slowly and deliberately, like someone with Parkinson's or cerebral palsy. He keeps wiping his chin to sop up the drool that escapes as he tries speak.

The woman who runs the house repeats what she thinks she has heard. "Dave, what do you think about what Jim said?" Dave, in a wheelchair, is missing a leg. Like Jim and the other four residents, he has severe medical problems. One of the men is terminally ill. Jim tries not to get close to anyone; the place is too transitional.

"You don't know what you're talking about," Dave mumbles.

Jim shrugs. It's easier than talking, and he has made his point. There's nothing to gain by arguing with this guy. These meetings are pretty much a waste of time. The phone rings in the kitchen and the director steps out to answer it. She calls Jim to the phone, which doesn't happen very often.

"Mr. Maki, it's Dr. Pomahac. I have some great news. I'd like you to come in right away. We found a donor."

"Really?"

"Yes, we did."

Pomahac hears Jim say something that sounds like "Then let's do it."

The meeting dissolves into a round of "Good luck" and warm wishes. The men are not close, but good news is hard to come by in this house. And not one of them would trade anything he's been through—and they have all been through a great deal—for a single day without a face.

The car service is called, and Jim gets in alone. At the start of the long drive into Boston they pass the donut shop where the guys go to buy pastries. If Jim could smile now, he would. After the operation he'll be able to eat whatever he likes. He can't wait to walk into that shop and buy something for everyone in the house. They all seem to like the lemon and raspberry donuts. He's always loved lemon, anything with the flavor of lemon, ever since he was a kid. He feels like a kid right now, on Christmas morning. A present is waiting at the hospital just for him, the biggest present anyone could ever receive. He can't believe it's really happening.

"This is Cynthia. Leave a message."

"Cindy, they found a face," he mumbles into her voice mail.

They've been married for thirty-one years, but they haven't lived together for almost twenty-five of them. Still, she is his wife. And there's Jessica, who is graduating from college next month. From college! She'll be excited that her father is getting a face. Maybe they will visit him in the hospital. He sure hopes so.

Butterflies are starting up in his stomach. He can't decide whether it's the motion of the car or the excitement he's feeling as they approach the hospital. Dr. Pomahac said to come right away, so they must be ready. Good. He hates waiting.

It's tough to be this excited and have nobody to talk with. He thinks about the things he might do after the surgery. Maybe he'll coach Little League. Maybe there's a kids' football league where he could help out. He'll probably need to get his bad eye fixed first. And he still can't use his right hand, so maybe coaching is out.

Enough of that. He won't think about all the things that are still wrong. Today is a day to focus on what's going right.

He swallows and leans his head back, trying to relax. What did he do to deserve this good fortune? Is it possible that his mother is somehow helping him from afar? He hopes she is watching out for him. Would she even recognize him with a new face?

chapter eight

Forty years earlier. April 20, 1962, midnight. Seattle.

Well-heeled citizens in black tie sip champagne from flutes with crystal stems in the shape of the new Space Needle, which is being christened tonight. They stand in the wind and the dark, straining to hear the opening remarks at the 1962 World's Fair.

A few miles away Jack Maki is just getting home after another long day at the university. The Phi Beta Kappa meeting dragged on, and as the presiding officer he stayed until the bitter end. Then he returned to his office to review his notes for tomorrow's senate meeting; next year he will become its president. He graded some of the student papers on his desk before finally heading for home.

After checking on John, he pokes his head into Jim's room. Fast asleep, both of them. Jim, who is twelve, is beginning to shoot up, his baby fat gone, his limbs lean and sinewy from many hours of basketball. Jack is proud of his talent, but he worries that sports are taking over the boy's life. It's time for more homework and fewer hoops.

He looks at Jim's peaceful face, his unbridled energy bottled up in sleep. He has been patient with his son, and Mary has been a saint. But Jim can't seem to behave. Jack cringes every time he walks through the front door and Mary tells him of yet another mishap. Yesterday it was a fight down the street.

"He called me a Jap."

"So? That boy is nothing to you."

"Dad, I had to do something."

"Yes. What you had to do was turn around, walk home, and get started on your homework."

"You don't understand."

"Don't I? Here's what I understand: if it weren't for education, we wouldn't be in this house. Education is what got your mother and me out of the camp during the war. Grandpa Ichi lost everything. They gave him a check for the farm, but it wasn't enough to buy a flower stand."

"They said my eyes were slanted. They said I squint."

"Grandpa is an old man, Jim. When he dies, he will have nothing to leave to his family. Can you imagine what that feels like?"

But as he stands above his sleeping child, he realizes that a twelve-year-old can't possibly know what that feels like. Even in the dark he can see a bruise on Jim's cheek from the fight. It's a cheek that looks a little exotic. Mary thinks Jim is part Native American, and that sounds right to Jack.

Jim apologized for fighting, but Jack held firm. "No baseball tomorrow. You're staying home. I don't want you outside. Do you understand?"

"Dad, no! The team needs me. They're gonna lose if I don't play. They'll hate me!"

"Go and get ready for bed."

• • •

It's always something with this kid. Last week it was five dollars that Mary left on the kitchen counter. "Mom said she was going to give me some money," he told his father. He turned to his mother and said, "I didn't think you would mind."

"Jim, you have to ask me first."

"If I catch you doing that again," his father said, "you'll be grounded. I don't care if you miss a game. Do you understand, son?"

"Yes, sir."

In Japan Jim once bought lunch for the entire fourth grade. The teacher, concerned about the amount of money in the boy's pocket, had called Mary. Then there was the street vendor who chased after him when Jim grabbed something off his cart on the way home from school.

His grades have been slipping. Junior high can be challenging, but Jim is smart. He picked up a lot of Japanese, and he knows everything about sports. Last week, in the back of the car, he gave an impromptu lecture on the migratory habits of polar bears. He's got a mind, that boy, if only he would use it. He needs to start high school on a sound footing. Maybe he should skip basketball next year to focus on his schoolwork.

Jim turns in his sleep, his long legs shifting under the covers. He will soon be taller than John, who is two years older. Things don't come as easily for John, but he tries hard in school. He is so quiet. Jack can't believe how two boys raised under the same roof can be so different.

He gently closes Jim's bedroom door. It's peaceful at night, the way a home should be.

"Dr. and Mrs. Maki," the principal begins, "your boy is what we call hyperactive. He has trouble sitting still."

"He's a very good athlete," Mary says. "He's also very smart."

"Fighting can't be tolerated in school, Mrs. Maki."

"But the other boy provoked him."

"We know what a good boy he is," says the principal. "But he needs to channel that energy. He can't go around getting into fights. I'm sending him home with you now. He can return on Monday."

Jack and Mary stand to leave. Shaking hands with the principal, Jack assures him it won't happen again.

"Dr. Maki, this really needs to be the last time."

"Yes, we know. Thank you."

Jim has been sitting outside the office. He has heard every word. At least he won't have to go to school tomorrow.

A few months earlier. Seattle.

"Jim, I just received a telephone call from a man who says he's a coach for a Gil Dobie League. Do you know anything about this?"

Jim has been sitting on his bed, pretending to do his homework as he listens to his father on the downstairs phone.

"He's putting a team together."

"Well, he's coming over to speak with me."

"Oh." Jim knew the coach was going to call. He tried out the other day, although he knew his dad would probably get angry. He wanted to see if he could make the team.

After tryouts he walked up to the chain-link fence where they posted the roster. Almost everyone made the team, but it was still a thrill to see his name up there: JAMES P. MAKI.

"Um, coach? I'm Jim Maki."

"I know who you are. What's the matter, son?"

"I'm not sure I can play. I wanted to see if I could make the team, but I don't think my dad will let me."

"Is he worried about injuries?"

"No, he's worried about my grades."

"Well, school's the most important thing, right? Are you having trouble passing your classes?"

"I can do okay when I want to. I get mostly Bs and Cs. My dad wants all A's and Bs."

"Do you want to play, Jim?"

Jim shrugged. "I guess so."

"Would it help if I talked to your dad?"

"Maybe."

It was worth a try. The kid was fast, he had great hands, and he could throw. Maybe a wide receiver, conceivably a quarterback. He could be college scholarship material. This wasn't the first time the coach had to make a house call.

"Hi, Mr. Maki, the name's Dick."

Jim and his father are stacking logs on the side of the house. Jack Maki looks up. "*Dr.* Maki," he says, smiling with a nod. "Jim, hand me another log."

The coach puts his hand back in his pocket. It is brisk, the air clear with a hint of fall. Fireplace weather. Football weather.

"You two need a hand?"

"No, we're fine." Jack stacks another log neatly on the pile.

"Dr. Maki, you're familiar with the Gil Dobie League?"

"No, not really. Football, isn't it?"

"Players are selected from around the city to be on teams. If they're good, they can join a senior league, the best from all over Seattle. Gil Dobie coached at U. Wash many years ago."

"I'm aware of that." Dr. Maki smiles. "I teach there."

"Yes, Jim told me. So you know he coached the Huskies to thirty-nine straight victories."

"That's interesting."

Interesting? The coach can see where this is heading. "Your son wants to play, and he's pretty good."

"Jim already plays Little League. That's sufficient. And he will not be permitted to play this spring if his grades don't improve."

"Dr. Maki, he's got it in him to be very good at football. He might be scholarship material."

"I'm a college professor. We have other avenues for scholarships."

"I can see your son playing quarterback if he works at it, Dr. Maki."

Quarterback! *Please, Dad.* Jim crosses his fingers behind his back, his palms sweaty.

"Jim needs to focus on his studies."

"Excuse me for saying so, sir, but it has been my observation that when a young man finds something that is very important to him, he will usually try to protect it. Jim says he can handle schoolwork when he tries."

"That's exactly right. When he tries."

The coach stands there quietly. Antagonizing the boy's father will get him nowhere.

"I'll hold Jim's spot for a week. Let me know if you change your mind."

He gets back in his car. Through his rearview mirror, he watches the boy run into the house. He can't quite make out if he is crying.

April 8, 2009, early morning. Hospital conference room.
Late last night Dr. Pomahac called his team members to tell them the good news. He swore them to secrecy, and most of them didn't sleep much after the call. Pomahac didn't sleep at all.

Now they have gathered in a very crowded conference room. Thirty-eight doctors, nurses, and technicians will attend the surgery. Pomahac reviews each step, checking everything one last time with each of them. The donor family has planned a bedside service with their rabbi at four o'clock this afternoon, and then they can begin. They expect the surgery to take more than sixteen hours.

Dr. Christine Kim, Jim Maki's hospital psychiatrist, is also there. Pomahac asked her to work with Jim almost as soon as he had him in mind for a face transplant. Even now the surgeon is haunted by what he saw the night Maki was carried into the Emergency Room after his face-first fall onto the electrified rail. He has shown Dr. Kim pictures of Maki that were taken after several rounds of plastic surgery. He didn't share the pictures he took that first night, when Maki came through the door. They were too horrible. So he asked his wife to sketch some black-and-white pencil drawings based on those photographs.

"I haven't mentioned a transplant to Mr. Maki yet," he told Dr. Kim the day she viewed the disfigured profile on the screen. "Would you meet with him and see what you think?"

Maki was guarded in their initial encounters. He hadn't requested a psychiatrist, and he wasn't sure why Dr. Kim was taking an interest in him. She seemed to be assessing him for something, but what? Years of living dangerously had taught him to be wary. Only later, when Dr. Pomahac asked if he would consider face transplant surgery, did he figure it out.

Her job was to determine whether Maki would be able to cope with what lay ahead, although nobody knew exactly what that might entail. He might be the first American to have a face transplant, with all the public scrutiny that would follow. For the past four years he has had virtually no contact with the outside world, let alone with members of the press, who will surely come calling if and when his identity becomes known after the surgery. He has been living in virtual seclusion, slowly recovering from the subway accident and the ten surgeries he has already endured. He is also a recovering drug addict. And he carries the emotional scars of a war gone wrong and, ever since the accident, physical scars as well.

Dr. Kim is well aware that war veterans sometimes have special issues. A couple of years ago she was urgently paged to attend to a patient who was having a major panic attack. He had fought in Korea and became hysterical when he heard a helicopter landing on the hospital roof. The loud thumping of the blades near his window brought back terrifying memories of the war. Now, as her colleagues drone on in their medical jargon about Maki's impending operation, she wonders again about his mental health. From all indications the medical team is ready. But what about Mr. Maki? She thinks he'll do well, but her psychiatric assessment is based as much on art as on science, and there is always room for an unexpected outcome.

In another part of the hospital, almost a city block away, two men arrive at Joseph Helfgot's room. Dr. Marcelo Suzuki is a professor of dentistry at Tufts University School of Dental Medicine, and with him is a young resident.

Suzuki, one of Maki's doctors, is a maxillofacial prosthodontist, a dentist who specializes in building parts of the face that stand in for a person's eyes, nose, or ears. Most of his patients are cancer survivors and burn victims. Dr. Pomahac has asked him to do something about the gaping hole he will leave when they remove Helfgot's face. Suzuki is here to design a mask, a mask made of silicone. He is a master craftsman, and many of his prosthetic facial devices can pass for the real thing. He's had complicated cases before, but this is completely unprecedented.

"Can I help you?" asks Lisa Kelley, the ICU nurse, as she adjusts a catheter.

"We're here from Plastics, to do some modeling."

"Of what?" she asks.

He is careful not to disclose too much. If she doesn't know about the transplant, it's not up to him to tell her.

He is barely listening as he stands in shock, gazing at Joseph Helfgot's face. The man has a beard! Pomahac didn't mention *that*.

He can't set a mold over a beard. The family will be coming in this afternoon for a final goodbye, so shaving it off is not an option.

He mumbles something about Vaseline as he and the resident start removing items from a black bag.

Lisa has been a nurse for a long time, and she knows when to back off and mind her own business. She leaves the men to their work, but she continues to watch them. She feels protective of her patient. She is also wondering what on earth these guys are up to.

Dr. Suzuki takes a large glob of Vaseline and begins to spread it over Helfgot's beard. Then he takes some kind of white casting material that looks like clay, puts it on top of the Vaseline, and works it to the sides before adding more to the cheeks and neck. When he is done, he and the resident stand around for a few minutes making small talk. He taps the white stuff a few times to check its consistency and then starts pulling the hardened material off Helfgot's face. It resists, sticking to the beard, and Lisa hears him mutter something under his breath. The resident is looking at her, so she turns to the computer screen, trying to appear engrossed in something else.

When they leave a few minutes later she sees that Helfgot's face is bright red where the clay has dried. But it doesn't seem to be clay. It's too shiny and too thin. Little white specks of it, together with Vaseline, are embedded in his beard. She puts some cream on her fingertip and massages it onto his cheekbone. She hopes Susan will get here before she has to go.

Lisa is leaving for Africa tonight, along with others from the hospital, including Kevin McWha, who has often nursed Joseph, and Dr. Rawn. It's their second year volunteering as a Brigham-based team at a heart clinic in Rwanda. Last year, when Joseph learned about the project, he helped the nurses pay for their travel expenses. She hopes she'll have a chance to say goodbye to Susan. She is sorry this is happening as they are leaving the country. She is sorry this is happening at all.

A doctor enters the room whom she recognizes from a surgical rotation a few years ago. She remembers him having a foreign accent. Dr. Pomahac nods to her and takes a quick look at Joseph. "We're good," he says to himself before he vanishes. What was *that* about? When she looks through the glass at a young physician, he gives her a *Don't ask me, I just work here* shrug.

April 8, 2009, 1 p.m. Emergency Room.
Jim Maki has been downstairs for a couple of hours in a special area of the Emergency Room. Everyone seems happy to see him. One of the nurses who took care of him during his most recent surgery stopped by with another nurse to help him pass the time. Since then it's been quiet. He'd like some lunch, but they won't let him have his usual thick shake, which seems unfair. Everyone around him looks busy, but he has nothing to do but wait.

Dr. Kim arrives. She is looking for signs of distress, but it's difficult with this patient, who has almost no face to observe.

"How are you doing, Mr. Maki?"

"Fine."

She marvels at his apparent calm. "Are you nervous?"

"Nope."

"Why not?"

"I just never am. I've been under the knife before. I'm ready to go." She believes him. This will be his eleventh operation under Pomahac's steady hand. "Do you know about the donor? Or how he died?" So he has been thinking about that. She wondered when that would happen. Recipients are almost invariably curious about the person who is giving them an organ, and their curiosity is typically accompanied by feelings of guilt. Somebody had to die to make this donation possible.

Kim has been reading up on transplant recipients ever since Pomahac asked her to work with Mr. Maki. She knows that after some time has elapsed, organ banks help facilitate an anonymous exchange of letters between recipients and the donor families. It is left up to the recipients to decide if they would like to become personally known to the family—assuming, of course, that the donor family wishes to be in touch. Kim can't imagine that happening here. Given the nature of the transplant, she is fairly certain the donor family will want to maintain their privacy.

All she knows is that the donor died in this hospital and that he was roughly the same age as Maki. Ever since the new HIPAA rules about privacy it is difficult to know anything about someone else's patient. And even if she did know something about the donor, she would not be allowed to tell Jim Maki.

"I don't know how he died," she tells him. "All I know is that he was a patient here and that he died a couple of days ago."

"Do you know how old he was?"

"Yes. He just turned sixty."

"I'll be sixty in September. But you don't know how he died?"

"No, I really don't."

"Was he married?"

"I don't know that either."

Maybe he too was in Vietnam. He wonders if their lives had been at all similar. He thinks this guy probably grew up in Boston.

"Do you know what he did for a living?"

"No. You know as much as I do. I realize how frustrating it must be not to know more, but the hospital is very strict about patient confidentiality. I'll stop by a little later. Try to stay relaxed, okay?"

Jim has one other question on his mind, but he doesn't ask it. Although Dr. Pomahac has repeatedly assured him that he will not resemble the donor, he can't help but wonder. This man who died, whose face he will be getting—what did he look like?

chapter nine

Winter 1538. The University of Padua, Venetian Republic.

Vesalius stands at a table, his hands covered in sticky blood. He is all of twenty-four, and the students who surround him are not much younger. Arriving from Leuven, Belgium, after a heated argument with his adviser over the best method of bloodletting to use on feverish patients, Vesalius so impressed the Padua professors that they soon awarded him a doctorate. Young men flock to his workroom, and on this late winter afternoon they strain to watch his surgical technique. As the winter twilight descends, he slices open the abdomen of a young female cadaver. Working quickly, before nightfall, is part of the challenge, part of what makes this exciting.

His most promising student, a young man named Gabriele Fallopio, stands opposite his mentor, assisting him and recovering one of the ovaries while Vesalius removes the other. Now, quickly, to the other room to preserve them in brine until tomorrow's light. Fallopio, who is obsessed with human anatomy, is fascinated by the mysterious origins of human life, and these walnut-shaped spheres, connected to thin, anonymous tubes, are part of the puzzle.

The students are excited. Tonight there will be a dinner to celebrate the private publication of the professor's anatomical drawings, and some of theirs too, in a small collection. Vesalius has arranged for a local bookbinder to put them together, and in a few hours they will finally see the finished volume.

The class has become much more lively since their teacher arrived. Vesalius likes it here because the Venetian Republic permits him to conduct his research on the human body without interference from the Church; almost anywhere else he wouldn't be allowed to work on cadavers. The students realize that his dissections represent a significant advance in the field, but they have no idea how much his work on human anatomy, and some of theirs as well, will influence the future of medicine.

Their drawings will become the basis for a groundbreaking work, *De humani corporis fabrica—The Fabric of the Human Body*. What will evolve into a seven-volume atlas will still be in print in the twenty-first century and will form the scientific basis for the field of developmental morphology. Centuries after its publication Charles Darwin will be heavily influenced by the work of Vesalius when he formulates his theory of natural selection.

April 2008. Experimental Morphology Lab at
Université Catholique, Leuven, Belgium.
Julian Pribaz, the director of Harvard's Plastic Surgery Training Program and a reconstructive microsurgeon at the Brigham, landed in Belgium early this morning with Bo Pomahac, his colleague and former student. They are jetlagged, but their adrenaline is high as they walk the narrow streets where Vesalius hurried along almost half a millennium ago. The visitors from Boston are here as guests of the head of the anatomy department, Dr. Benoît Lengelé. Lengelé travels frequently to France, where he collabo-

rates with Dr. Bernard Devauchelle and his colleague Dr. Jean-Michel "Max" Dubernard from Lyon, who stunned the world with the first successful hand transplant in 1998.

Several years later, in 2005, Dubernard shocked the world again with the news that he had transplanted a human face. This announcement was not well received. Many doctors denounced the transplant as nothing more than a stunt. They believed the medical risks did not justify an operation that was not lifesaving.

But not everybody felt that way, especially Dr. Joseph E. Murray, who performed the world's first successful organ transplant in 1954, a kidney donated by a living donor to his identical twin brother. The double surgery that made Joseph Murray famous was performed in Boston at what was then known as the Peter Bent Brigham Hospital. It took a while for the judges in Stockholm to get around to it, but in 1990 they awarded Murray and his colleague, E. Donnall Thomas, the Nobel Prize for Medicine.

As a young physician during World War II, Joseph Murray had worked at Valley Forge Hospital in Pennsylvania, the country's leading burn center. Some soldiers arrived with such extensive burns that skin grafts normally taken from other parts of their body were not an option; there was simply no skin left to take. Instead Dr. Murray and his colleagues would graft skin from fresh cadavers. But sooner or later the soldier's body would reject the foreign skin. Still, these grafts bought precious time for the soldiers' wounds to heal from within, and for other parts of the body to recover enough to allow for an autograft, where rejection was not an issue.

Murray had trained at Harvard Medical School, and after the war he returned to one of its teaching hospitals. His desire to solve the fundamental problem of rejection led to an association with Peter Bent Brigham Hospital's Surgical Research Laboratory, which was established in 1912 by Dr. Harvey Cushing, the father of modern neurosurgery. As the Brigham's new surgeon in chief, Cushing insisted that his lab be housed in the Harvard

Medical School's quadrangle around the corner, where he was able to use animal specimens and introduce students to the latest surgical techniques.

Years later it was there, under the guidance of more experienced physicians, that Murray and his colleagues successfully transplanted kidneys into dogs. Eventually they were able to carry out the first successful human kidney transplant between identical human twins, born from a single egg that split in the womb. They shared duplicate DNA, making rejection virtually impossible, as each brother's immune system recognized the other as itself. In 1954 the kind of DNA testing that would demonstrate that the twins shared the same genetic material did not yet exist; DNA's chemical structure had been discovered only a few months earlier. Instead, the young men's fingerprints were analyzed at a nearby police station. They were similar in every way, much more so than even fraternal twins.

Half a century later Dr. Murray was thrilled when he heard about the first face transplant. A few days later he attended a meeting with Elof Eriksson, his successor as chief plastic surgeon at the Brigham. Referring to the naysayers, Murray said, "Elof, this controversy is exactly the same one I went through after the first kidney transplant." Back then a research fellow who had wanted to study with Murray was warned off by his chief, who dismissed Murray and his colleagues as "a bunch of fools." And a Harvard professor who was one of Murray's closest friends took him aside and said, "Joe, please don't get involved in this. It will never work, and it could ruin your career."

Joseph Murray was frustrated and disappointed by the negative reaction that Max Dubernard was getting. He knew that face transplants could repair broken lives, but there was a more personal reason for his dismay. Forty years earlier Dubernard had been Murray's student. Murray liked the young, chain-smoking French doctor who had come to Boston in the mid-1960s to study under his guidance. They became close, and by the time Dubernard returned to France he had learned a great deal about human organ transplantation. It was said that he had also left behind a few broken hearts.

As he reminisced with Eriksson about his newly famous protégé, Murray asked, "So when do you think *we'll* jump in?"

"We'll get there," Eriksson replied. He ran through a mental list of the young surgeons rising up at the Brigham. They all had talent, but talent wasn't the issue. In 2005 the United States wasn't ready for face transplants. They needed a young doctor with the necessary fire to bring this advanced surgery to the Brigham.

The doctor who best fit Eriksson's description was Bohdan Pomahac, but Eriksson didn't know that yet. Neither did Pomahac—not until 2008, when he sat in a lecture hall and listened to a presentation by Max Dubernard. Eriksson had just established a visiting lecturer series in honor of Joseph Murray, and Dubernard was given the honor of inaugurating the event.

Dubernard told the audience that he had been thinking about facial transplantation ever since he had performed the first hand transplant. He imagined a single surgery that would encompass the transplantation of part or all of a human face: muscles, nerves, blood vessels, bone, and skin. He conceded that enormous advances had been made in plastic reconstructive surgery, but even after multiple operations the most severely disfigured patients were left with woefully inadequate results.

For some of these people war injuries had shattered their jaws and cheeks. Others survived blazing infernos with their skin and nerve endings burned away, which required weeks of debridement to remove slowly decaying flesh. Still others had been badly mauled by animals or attacked by savage tumors.

Lifting even part of a donor's face with millimetric precision and fitting it perfectly onto another person was close to miraculous. And as with any transplant, the recipients would have to take immunosuppressant medications for the rest of their lives. These drugs have serious side effects, which can be deadly. Although skin is an organ, along with the heart and the kidneys, some questioned whether a face is really necessary for survival. Was it worth the risks that such a transplant would entail?

Dubernard told the audience that his own doubts were erased when, at Dr. Devauchelle's request, he visited a woman named Isabelle Dinoire, who had survived a brutal attack by her dog after she overdosed on sleeping pills. She had fallen and hit her head, and some people believed the dog had been trying to wake her up. Whatever the reason, the savage encounter left the poor woman without a nose, lips, right cheek, and chin.

After one look at Dinoire, Max Dubernard jumped into action, taking Devauchelle with him. He thought about how he would feel if Isabelle were his daughter. This woman needed help, and these two doctors were going to do everything they could to provide it.

They began to treat her wounds in a new way, forgoing the usual repair regimen of one surgery after another. Instead they focused on intensive physical therapy to keep her face muscles strong and scar tissue at a minimum as they waited for approval from the French authorities. While the committee considered the question of a face transplant, the French surgeons performed facial dissections on cadavers, practicing over and over again. Practice and more practice, they believed, was the key to success in the operating room.

Pomahac sat there transfixed as he listened to Dubernard. He had first discussed facial allograft with Dr. Eriksson soon after the French team performed the groundbreaking surgery in 2005. He even began to formulate a protocol for face transplants at the Brigham, with the idea that they would be offered to patients who were organ recipients and already on lifelong immunosupressant drug therapy, so a face transplant wouldn't subject them to any additional risk. Pomahac received preliminary approval in 2007, but nothing had come of it because he didn't have any patients who had already received an organ transplant and who also needed a face. How many such people could there be?

Now, as he sat in the lecture hall listening to Dubernard, he thought about several of his patients, including Jim Maki. They needed faces, desperately. By the time the lecture was over Pomahac knew this was something he was going to push hard for. There was simply nothing, *nothing* that could help these patients short of a face transplant. He had to try.

After the lecture Eriksson introduced him to Dubernard, who told Pomahac and Pribaz, "You guys have to do this," adding that his door was open whenever they were ready. Dubernard understood, just as Joseph Murray had half a century earlier, that only two things would quiet the scathing criticism he had received for the Dinoire surgery: time and more facial transplants. Who better to make that happen than Murray's successors?

Back in 2005 American medicine was not ready for face transplants. After Isabelle Dinoire's surgery, the American Society for Reconstructive Microsurgery released a position paper opposing them. In their opinion the potential for rejection, coupled with increased risks of infection, cancer, and diabetes from immunosuppressant drugs, far outweighed any physical improvement that a face transplant might provide.

And anyway, how important were those improvements? "Most patients with facial deformity adapt quite well and accept their physical appearance as 'self,'" the authors of the paper argued. "The psychologic repercussions of a facial transplant on family and friends of both donor and recipient cannot be underestimated. The ethics of inflicting an untried, and potentially fatal or deforming remedy for the purposes of advancing science must be carefully weighed against the Hippocratic credo of doing no harm."

Bo Pomahac, himself a member of the society, understood these reservations, and for the most part he shared them. But his focus was on extreme cases, patients whose problems went well beyond disfigurement, who often needed a feeding tube to eat and a tracheostomy tube to breathe. How they looked was the least of their problems.

It took a while, but when Pomahac finally received approval from the Brigham's review board, he took Dubernard at his word and made travel plans.

He and Pribaz flew to Brussels last night and now sit across a table from Devauchelle and Dubernard. The review board and their colleagues at home have agreed that James Maki is the person most in need of a face transplant, and the Boston doctors have brought his entire medical history with them.

They discuss Maki's case, and the French team agrees that he is the logical choice. For the next three days, under the watchful eyes of the men who first performed this groundbreaking surgery, Pomahac and Pribaz practice on cadavers in Dr. Lengelé's lab. Pomahac is impressed by Devauchelle's surgical technique. He is a skilled guide, and Dubernard is a brilliant communicator. By the fourth cadaver the Americans have reduced the time it takes them to recover a face from four hours to an hour and a half.

Before they fly back to Boston, Pribaz and Pomahac are taken to dinner by their hosts. Incredibly this is the first time the entire French team has been back together since the face transplant they performed in 2005. They have brought along a surprise guest: their patient, Isabelle Dinoire.

She looks fantastic, Pomahac thinks. He watches her eat, savoring each bite without difficulty. Although she speaks softly, her voice is clear and strong. She dines easily, talking and chewing and sipping water without any apparent effort, and laughs at jokes. She even slips outside with one of her surgeons to enjoy a quick cigarette. Now *that* would make a hell of a picture, Pomahac thinks. But of course the French do things differently.

"I have a lot to do when we get back," he tells Pribaz on the plane back to Boston. As it turns out, he has exactly one year.

chapter ten

June 30, 2005. Malden, Massachusetts.

for the past few months James Maki has been living in a half-way house on Boston's North Shore. As a former heroin user he's been on methadone for years, but he still scores other drugs, mostly pills, which are easy to find on the streets of the city. He pretends to be sober and is very good at fooling people, which is one reason he has survived for so long. As long as he doesn't bring drugs into the house, they can't throw him out. And he can't afford to get thrown out because he has no place left to go.

Cynthia has long since washed her hands of him. When their daughter, Jessica, was born in 1985, he made all kinds of promises that the past was the past. He was in jail the day Jessie was born, after a policeman found a large amount of marijuana in the trunk of his car. Cindy was with him, pregnant with Jessie. When he was sent away, his lawyer took Cynthia under her wing. They were friends from college, and the lawyer persuaded her to take the LSATs, the law school admission test, and perhaps go on to law school. Cynthia didn't do well on the test, but she went back to school and became a paralegal.

On the day Jessica was born they let Jim out of jail to see his new daughter. The prison warden had coached basketball at a western Massachusetts high school and remembered Jim as a gifted player. He knew he wasn't dangerous, merely troubled. A police officer dropped him at the hospital, promising to return in a few hours. When Cynthia gave him the baby to hold, he kissed her tiny forehead and took in her baby smell. He couldn't believe they had created such a beautiful child.

Cynthia got her own apartment, and Jim showed up the day he got out. He promised there would be no more drugs and that he would stay on his methadone treatment and clean up his act. He said he would find a real job rather than picking up the odd painting or housecleaning gig like so many of the users he knew.

He was so glad to be out of that hellhole. The guards treated him as if he were stupid, as if he were nothing more than dirt. If you took more sugar packets than you were supposed to, they called it stealing.

On his first night back with Cindy he went out to celebrate his new freedom. He didn't return until the following night, high as a kite. Although she was furious, Cynthia couldn't bring herself to tell him to leave. He would come and go, and she let him hang around when he wanted to.

One evening when Jessie was in preschool a social worker showed up at their door. A neighbor had called, concerned about some of the activities at Cynthia's apartment. Jim's mother was very sick, and Cynthia's mother was seriously depressed, so Jessie stayed with a friend while Cynthia got some counseling. The social worker gave her an ultimatum: If Jim didn't move out, they would take away her daughter.

Jim or Jessica. That was a tough year for Cynthia. Her mother-

in-law, the one person who had always been there for her, was
completely bedridden, slowly dying of rheumatoid arthritis that
was attacking all of her bones. Cynthia felt alone, vulnerable, in
a way she never had before.

Mary Maki was a love, kind and sweet and gentle. On week-
ends Cynthia would bring Jessica to spend the day with her
grandparents. When Jessie heard "We're going to Grandma's
today" the little girl's eyes would light up in delight. They would
drive to Amherst, in the western part of the state. Dr. Maki had
been teaching at the University of Massachusetts since Jim was a
senior in high school and had built a Japanese-style home.

Jessica would walk around the property with her grandmother,
exploring its delights. They would pick flowers together, and Mary
would sit with the little girl and arrange them, along with the twigs
and moss that Jessie found, into small masterpieces of design. Jessie would present the arrangement to her grandpa, and she and
Jack would play cards together while Cynthia and Mary sat in the
kitchen as the rice steamed. But then Mary began to fade.

Jim came to see her shortly before her death. He was not allowed to visit his wife and daughter, and he missed them terribly,
especially now that his mother was dying. He had never felt so
low in his life.

His father and a hospice nurse were there when Jim arrived.
"Don't say a word about Jessie having to stay with a friend," Jack
told Jim. "Your mother is already worried sick about what will
become of you after she's gone, and I don't want her to start worrying about Cynthia and Jessie."

But Jim couldn't help himself. His mother had always comforted him when things were bad. She knew what to say to make
things better. Jim told her everything, and after he left, Jack found
her in tears. She died soon after that visit, and Jack blamed his
son for hastening her death. He never forgave him, but he did
write Jim a check after her funeral. Jim was entering a new half-

way house and needed some spending money. "But if any of this goes to drugs," Jack said, "there won't be any more, ever. This is the very last time I'm going to bail you out."

A few days later the woman who ran the halfway house called Dr. Maki. He had asked her to let him know if Jim was looking for money, and apparently he had asked for a loan. Jack had given him two hundred dollars, more than enough to cover incidentals like toothpaste and shampoo. The professor was true to his word; he never gave his son any more money. When Jim called he would grunt a few words and hang up.

After his mother's death Jim moved around from place to place, dropping in on Cindy intermittently when he needed food or a little money, or a place to take a shower and sleep for a day. She was never able to turn him away. He was, after all, her husband and Jessie's father. And he didn't have a mean bone in his body.

She blamed Vietnam for his troubles. That was where he first tried heroin and where he saw some terrible things. She sometimes thought there was a demon inside him.

Once he was so desperate for a fix that he threw a television out of his parents' bedroom window. He was sick with need, but Jack refused to give him money. The television landed on a solar panel below, shattering the glass. Jim was so apologetic afterward. "You don't know the grip heroin has on me," he said. "I'm sorry it happened, but there's no rectifying it." Not much later he entered rehab yet again. No matter how many times he cleaned up, though, Jim and trouble always managed to find each other.

In 1999, while visiting Cindy, he came out of the shower and put on clean clothes. He walked a few feet and collapsed, straight to the ground. Jessica was thirteen, and she watched in horror as her father lay there, convulsing.

"Jessica!" her mother screamed. "Go outside and stay there!" An ambulance rushed Jim to the hospital. He had suffered a brain aneurysm that should have killed him then and there. Thank God he'd been at Cynthia's. It was touch and go for several days, and he struggled for a long time to get back on his feet. But even after another lease on life, he soon turned back to drugs. That was the last straw for Cynthia. She was done with him, completely done.

Now, on this fine summer morning in 2005, Jim has jumped on the train from Malden heading into Boston. It's easy to score in the city, and he has a little money from working for a guy who pays him to clean houses. This guy's no dummy. Everyone he hires uses drugs, and he knows it. He pays them first thing in the morning, rather than after they've finished working. That way they can go off and score, and come back and work like crazy men, not even stopping for lunch. If they don't return to work, they won't get paid the next morning, which means they won't get their fix. He's got it all figured out, this guy.

It's one of those days you hope will never end—not too hot, but nicely warm with a gentle breeze. New Englanders wait months for a day like this. Night has fallen, and Jim and a few buddies are hanging out in Southwest Corridor Park, a five-mile strip of grass and trees laid down when the old Orange Line tracks were sunk underground a decade earlier. There's a pickup basketball game under the lights, and Jim watches the young men running up and down the court. He's jealous of their youth, but he enjoys watching them play. It's late, and he should probably be getting on a train back to Malden. It could be after midnight, but he's not sure. He is feeling no pain. Time is moving slowly, like it always does when he is high. He is dizzy and a bit shaky, but he'll be all right. He'll sit here until the dizziness fades, and then he'll get on a train home. The Ruggles stop is nearby. A lot of people are still out, walking around, sitting on park benches, trying to extend this beautiful evening. There's no rush. He'll just sit here a while longer.

• • •

A few hours later. Trauma and Burn Unit.
Bo Pomahac is on call, dozing in a chair. He's exhausted. Having completed his surgery program just last year he gets the tough shifts. Hanka and Bo had their first child, a girl, barely two years ago, and now they have a baby boy. The little guy is one now and is trying to pull himself up to a standing position. Their daughter is a moving ball of energy. How Hanka keeps up with them all day he'll never know.

He is startled out of sleep by his beeper calling: a burn victim, somehow electrocuted, is coming in. The paramedic in the ambulance is briefing the Emergency Room staff on the speakerphone as Pomahac rushes in. A man has fallen face-first onto an electrified subway rail at the Ruggles station. Miraculously he still has a pulse.

They crowd around the speakerphone. The paramedic is saying he's never seen anything like this.

"How bad is it?" asks one of the doctors.

"Burns. The middle of his face, or what's left of it, it's actually kind of gone. His right arm is really burned up, especially his hand. Jeez, he's a mess."

"Is he breathing?"

"Intubated."

They hear the wail of the siren approaching outside. The paramedic keeps talking as he and his partner bring the man in on a gurney. The ER team brace themselves, preparing for the worst as they try to picture what they have just heard. But it's impossible. What does a man without a face look like?

The gurney bursts through the door. Pomahac stands back, behind the triage team. He takes one look at the man and he's stunned. He has to do something, but what? He has never seen anything like this. The man has no nose, no upper lip, and no facial skin. Pomahac clamps his teeth together, the adrenaline rushing through him as he fights off waves of revulsion.

"There's no ID on him."

"I don't think he's going to make it," says one of the docs.

"Get Social Services down here. We've got to find his family."

"We're losing him!" Without a family member to stop them, they resuscitate. But nobody believes he will last the night.

chapter eleven

Susan is making final preparations for Joseph's funeral, which will take place Friday morning. It's a little tense as both the temple and the funeral home try to accommodate the family. She called them on Monday, and Jewish law encourages a timely burial. But relatives are still arriving from abroad; Joseph's daughter, Emily, has just come from Paris, and a cousin is flying in from Israel. The Helfgots are not a big family. Joseph's mother, father, and sister are all dead, and almost all their other relatives perished in the Holocaust. Joseph's adult nephew, Bobby, is playing cards with Jacob at the kitchen table. Bobby looks lost. Uncle Joseph was his stand-in father, and his mother, Joseph's sister, is dead.

Passover starts tonight, and funerals are not normally held on major Jewish holidays. The director of the funeral home has been in touch with the secular Independent Workmen's Circle Cemetery, a fitting resting place for this hardworking child of immigrants. Technically Joseph too was an immigrant; as a child he came through Ellis Island with his parents. The cemetery officials have agreed to bury him on the second day of Passover, if they can find the space.

A delivery van pulls up to the house with food, flowers, and a case of wine from Joseph's friends at HBO in New York. The driver brings everything into the kitchen and hands it off to Susan's girlfriends from Los Angeles. "Sue, where should we put this?"

I don't give a damn where you put it. "Oh, anywhere is fine. How about taking the wine out to the garage? And I think there's a vase on the top shelf."

Everyone is sending over boxes of matzoh. We should be making a Seder, not planning a shiva. Joseph, do you remember when we were buying this house? After the Realtor told us about the heating and the roof and the plumbing, you asked, "Suze, how many can we squeeze in here for the Seder?"

Oh, sweetie, your Haggadah. I forgot all about it. The one you just finished making. Everyone with their assigned parts, all thirty-four of them. Even the man who delivered the rug two weeks ago! Did anyone tell our guests not to come tomorrow for the second night? I guess they all know you are dead. Well, maybe not the rug man. I'm glad I didn't take your Haggadah to Kinko's to make copies. I was going to, but then Dr. Lewis called and we went to the hospital. Oh no, the food!

"Ben, I need you to run over to the Butcherie. I think they're still open." They'd better be. A large turkey and an eight-pound brisket are waiting to be picked up to supplement the gefilte fish, the homemade chicken soup with matzoh balls, the chicken breasts, two whole salmons, vegetable tsimmes, artichokes, the spinach soufflé, and various other dishes for a group of almost three dozen participants at a wonderful Seder that won't take place this year.

Susan must attend to a less pleasant task. Accompanied by her two adult stepchildren, she walks over to Stanetsky, the Jewish funeral home on Beacon Street. The air is heavy with moisture, but at least the rain has stopped and the day is mild. It feels good to be outside and away from all those well-meaning but inane questions about what should go where.

She is carrying one of Joseph's suits in a garment bag. They bought it last year for Jacob's bar mitzvah and had the tailor put in special panels to hold the five-pound batteries that kept the VAD machine going during the service. The alterations were surprisingly expensive. "*Gonif*," Joseph said when the tailor told him what the panels were going to cost. "I'll give you half, or forget it."

"Sweetie, don't start," Susan said as Joseph sat there with an oxygen cannula up his nose. "I don't have time to be sewing pockets in suits." In the end the tailor knocked off a few dollars. *Gonif*, Susan thought as she drove them home.

This is so weird. Four years ago I walked my mother's clothes down the street in the other direction, to the Irish funeral home. It's closer than Stanetsky. Come on, Susan, what does that matter?

"What day is it?"
"I think Wednesday," says Emily.
"What's the date?"
"It's the eighth," says Jon.

They declared him dead yesterday, April 7. He really died Sunday, during surgery, but April 7 is on the death certificate. Mom died on the seventh. Joseph had just gotten out of the hospital and was home in bed. I flew Mom to South Carolina to be buried next to Dad. Hardly anyone came to her funeral. Everyone is coming to this one. I don't think I can do this.

They enter the funeral home. Susan hates this place, which brings back painful memories of other funerals she has been to, like the one a while back for Sol Levine, Pam's father. Raised as a devout Catholic, she worries whether she'll know what to say to the funeral director. The logistics of this death have not been easy, and she sensed some discomfort on the phone when she spoke to him, especially when she explained that Joseph's body won't be released from the hospital until tomorrow afternoon. And Passover starts at sundown tonight.

"I hate this," she says.

"We'll do it fast," Jonathan assures her.

The funeral director ushers them into his office. Emily slumps into a chair, jet-lagged and still in shock. A riot of soft honey-red curls surrounds her face, which is pinched from so much crying. She swoops it up and stuffs it into a scrunchie.

The director tries to appear relaxed, but he must be eager to get home to prepare for his Seder. They review the wording for the obituary. Then Susan leaves with him to pick out a casket. Emily and Jon can't bring themselves to go with her.

It has to be cherry, like my mother's coffin. He loves cherries—no, I guess that should be loved cherries—more than anything, except maybe watermelon. God, he loved watermelon. This is, what, my fifth casket in ten years? I should have bought them in bulk! Joseph, his sister, his mother, and my parents—they're all gone so soon.

"How much is the cherry?"

The funeral director describes the silk lining and the Star of David. "It's a bit steep," he cautions.

"Whatever. The lining . . . I don't care, it's fine. I'll take it."

She stands in a daze, looking around at a room full of caskets.

God, please don't make me come back here ever again.

As they head back to the director's office Susan is barely able to hold herself together. "I picked out one just like Grandma Rachel's," she tells her stepchildren, "except this one's cherry, like my mother's. Grandma's and Auntie Pauline's were oak." They don't remember. Young people don't dwell on those kinds of details.

The funeral director clears his throat. "I have good news. It took a while, but we found a double plot at Workmen's Circle in West Roxbury. They're the only ones who have a double available, and they're willing to do the funeral on Friday." He pulls out a map indicating where the plot is located. He is relieved that it all worked out. He quotes a price and folds his hands on the desk. He looks at Susan, waiting for her response.

"Excuse me, did you say a *double*?"

"Yes, for you and your husband to be buried together."

"But I'm not Jewish." She winces as she says it. She knows that only Jews can be buried in a Jewish cemetery.

"You're not?" He is incredulous. Because she has shown so much familiarity with Jewish death rituals, he naturally assumed . . .

"No, I was born Catholic."

"I see." But he doesn't. He looks down at his papers, suddenly at a loss. All this careful planning, and she's not Jewish? He glances at the clock on the wall. It's almost three. Sundown isn't that far off. No one at the cemetery office will still be there.

"You know," says Jonathan, "my dad was kind of a big guy. I mean, he lived big. He always flew first class. He liked a lot of space."

The director doesn't follow.

"We'll take the double. Put him in the middle."

He nods. "Do you have the clothing for the deceased?" Susan hands him the bag.

"Smooth," Emily tells her brother as they walk out the door. "Really smooth." They all crack up laughing.

Joseph, are you listening? Jon is so fast and so funny, just like you. And so strong. I didn't know he was that strong. I could never have given Jacob the news without Jon by my side. And Emily misses you so. She looks so much like her grandmother and can size people up just as fast. She's so sharp. Whatever are we going to do now?

April 8, 2009, mid-afternoon. Department of Plastic Surgery.
Dr. Pomahac is meeting with his team one last time. The surgery is now scheduled for 6 p.m., right after the Helfgots say their final goodbyes. Pomahac spent some time this afternoon visiting with Jim Maki, who is eager to get started. Then he had a quick talk with Peter Brown from Public Affairs.

"They want to put a wire on you," Brown said, as though it were the most natural thing in the world. "The cameras will be filming throughout." For several months, a television production team has been filming *Boston Med*. Now, a few weeks before they are scheduled to leave, they have been invited to capture this historic surgery. Nobody expected they would find a donor before the crew left.

"You want me to wear it the whole time?"

"Ten years from now we will all be very happy we caught this on tape. And you can turn off the mike whenever you want."

Pomahac remembers a TV interview that Brown arranged for him a few years ago about facial transplant surgery. The resulting publicity became the grease for the slow wheel of approval for this project. Peter has never steered him wrong.

"Okay. You're the boss."

At the final team meeting Pomahac runs down his checklist one more time. Dr. Elof Eriksson is content to sit and listen, to do whatever it takes for this young man to be satisfied that nothing has been left to chance. At this point, he knows, the only thing left to do is to get the donor and the recipient into their respective operating rooms so they can start. Eriksson has been head of plastic surgery since 1986, when he took over from Dr.

Murray. Although he's been part of several breakthrough moments, he has never seen a group of medical people as invigorated by their task as this team is today.

Most of the credit belongs to his protégé, Bo Pomahac. There have been many administrative hurdles. Before they finally gave their approval, the review board came back time after time, requesting certain changes. The finance department had to waive the cost of the operation, and the medical team had to donate their services. He is proud of what Bo has been able to achieve. It was not easy. He knows that Joe Murray is proud as well.

Behind Pomahac's quiet demeanor is a determined soul, as Eriksson discovered a decade earlier, when the twenty-five-year-old sent him a letter requesting an interview. Pomahac, who was about to graduate from medical school in the Czech Republic, wanted to continue his education in the United States. He didn't know anything about Eriksson other than his name. Because it sounded European, Pomahac figured that Elof Eriksson might be willing to meet with a young visitor from central Europe. It was a good guess.

Thirteen years earlier. Brigham and Women's Hospital.
"We have nothing open at the moment," Dr. Eriksson reminds the earnest young man who has come to see him. He had told him the same thing on the telephone. Pomahac has just arrived in Boston, two days after earning his diploma from the Palacky University School of Medicine in the Czech Republic.

"I was hoping you might find something for me to do. I'll do anything you need."

Dr. Eriksson is impressed by the young man's sincerity and pluck. "We've got a lot going on here," he says as they step into Eriksson's Tissue Repair and Gene Therapy Research Lab. Soon they hope to transplant healthy cells directly into patients' wounds to encourage healing. Before long this lab will be a nurs-

ery for breakthrough gene therapies. Researchers will soon ma-
nipulate viruses to carry genetic material that can invade and
disrupt cancerous cells while leaving normal cells unchanged.

Pomahac is dazzled, although he isn't sure exactly what
they're really doing here with all this measuring and pipetting
and centrifuging. Small electronic monitors are plugged into almost every outlet, and people in lab coats scurry around in a controlled but chaotic-looking dance. He knows only that this is the
Boston he dreamed about. In the school where he was trained
there was no money for even the most basic lab equipment. They
read dated journal articles because nobody could afford a current
subscription to a major medical publication. His lab training, he
sees now, was a joke. Most people with Pomahac's level of prepa-
ration for a place like this would turn around and walk out the
door. But his resolve runs deep.

"Stay and watch for as long as you want," Eriksson tells him.
"Stop in before you leave."

At the end of the day Pomahac thanks his host and surprises
him by asking, "Do you think I could stay?"

"Stay?"

"I mean, could I come back again tomorrow?"

A few months later Eriksson has managed to scrape together
some money for a small stipend. Bo has quickly exhausted his
modest savings from working on a hops farm back home. He was
a laborer in the bagging station, picking tiny pinecones from female plants to make beer, filling fifty-kilo sacks and hauling them
to a grinding station before returning to fill yet another sack.
It started as a mandatory government summer job for students
while he was in medical school.

He met his girlfriend, Hanka, on the farm. She was a medical student at the same school. They have been dating for several
years now, e-mailing over CompuServe every day and talking on
the phone once a week. They met in 1989, just before the Velvet

Revolution transformed their country and the Berlin Wall fell. By the following summer Czech farms could no longer count on free labor. Pomahac helped a friend organize groups of students to harvest the crop, and he saved enough money to visit the United States. This is his second trip. In 1992, as a foreign exchange student, he flipped hamburgers on the New Jersey shore.

His parents used their savings for the plane ticket that brought him to Dr. Eriksson's lab. But Boston is an expensive city, and the three thousand dollars that he has managed to save won't last very long. Dr. Eriksson has come up with a few dollars for his work in the lab; although it's a tiny amount, it makes all the difference—as long as Pomahac sleeps on a friend's couch.

Eriksson is impressed with his progress. He is a quick study, and he is determined to make the most of this opportunity. Pomahac has much to prove. His father's promising career as a chemical engineer was cut short when he signed on with the anti-Communist uprising in 1968, the short-lived Prague Spring. When Russian tanks rolled in and squelched the revolution, his father was blacklisted and forced into a series of low-level jobs. His mother, a schoolteacher, joined the Communist Party so their two sons would be allowed to attend college. That helped, but not completely. After finishing his grueling course work to become a pilot, his brother was informed at the last minute that he would not be permitted to join the air force. Bo Pomahac took notice. If he was going to be successful it probably wasn't going to happen in the coal-mining town of Ostrava. Now that he has a foothold in Dr. Eriksson's lab he is not about to let go.

April 8, 2009. Department of Plastic Surgery.
The human body is a collection of microscopic spheres, each one holding a blueprint copy of the mystery of life. Each tiny cell is bound together by a membrane of fatty lipids that crowd to-

gether like bubbles in a bathtub, pierced by proteins and sugars. Take just the right mix of a hundred trillion or so of these cells, and you have a human biome.

Tonight the doctors will dismantle a fragment of this micro-scopic universe from one human being and reassemble it onto another, one tiny piece at a time.

We have come so far, Dr. Eriksson muses. Look at the disciplines represented here: anesthesiology, infectious diseases, pathology, psychiatry, all the surgical specialties, and more. The more complex medicine becomes, the harder it will be to make moments like this happen. Half a century ago, in Joe Murray's day, the big obstacle to experimentation, and to innovation, was the fear of looking foolish to your peers. So much has changed, including committees and protocols and government regula-tions, all of which slow everything down but add valuable cir-cumspection. When Murray operated on the Herrick twins in 1954 almost nobody owned a television. This surgery is about to be filmed for a TV series. Add runaway health costs to the mix, and it's a wonder anything experimental happens these days.

And yet, in spite of everything, they have made it to this mo-ment. Maybe the miracle isn't the face transplant itself, but the fact that they have been able to cut through the red tape to per-form one at all. And the main reason is the quiet, steadfast deter-mination that Eriksson spotted over a decade ago in his young Czech visitor.

He gives his protégé a reassuring nod. It's time they got started.

chapter twelve

eturning from the funeral home Susan finds large branches of cherry blossoms spilling from vases in every room. The house has been transformed into a park in spring.

"It's so beautiful," she tells the friend who has been helping her prepare for the crush of visitors they expect after the funeral. The woman is a Boston restaurateur who often travels to Manhattan to spy out the next trend, sometimes taking her daughter and Susan's son, Ben, who have known each other for years. When Joseph was healthy he would tag along too. His business meetings often coincided with these trips, and even when they didn't, he loved going to restaurants, especially in New York.

Susan used to go on similar outings in the early 1980s, when she was involved with a restaurant on Beacon Hill. Joseph was a patron, which was more or less how they met. Now their son, Ben, is interested in the restaurant business—Ben, who has done so well on the science and math AP tests, who has the talent to be a doctor or a scientist. Susan herself never listened to her parents' advice, and Ben seems to be taking after his mother.

Twenty-eight years earlier. Back Bay, Boston.
Susan steps into moist air as the city moves toward a lazy mugginess that threatens afternoon thunderstorms. Her third power breakfast at the Ritz in six months with the same lawyer seems to have finally paid off. The names of two prospects are written on his business card, which jostles beside the lipstick in her small clutch purse. When she gets back to the office she will call them and try to schedule appointments.

As she steps off the elevator the receptionist says, "Some man called you twice. He seems anxious to speak with you." She hands Susan several pink message slips. Two of them are from Joseph Helfgot.

The night before, Susan sat on a bar stool at Jason's, a popular downtown restaurant that would later become a Hard Rock Café. She and a partner from the restaurant were at the bar, pitching a woman from a convention bureau that sponsored fall foliage trips to Boston, trying to persuade her to make their Beacon Hill restaurant a lunch stop on the bus tour. She seemed to be receptive.

Three young men sidled up to the bar, all of them drinking beer in bottles. They wore khakis with loafers and slightly different hues of blue oxford shirts. They had loosened their ties and looked like new recruits for a security firm or insurance company. One of them tried to buy her a drink. "Tell me," she asked him, "do you guys always go out dressed alike?"

A man sitting at the bar laughed. Susan recognized him as someone who had been in their restaurant a few times. She was twenty-seven; he looked to be a little older. *He's definitely Jewish,* she thought, taking in that full head of curly brown hair and the beard. He was thin, with piercing blue eyes.

When the young men moved on, the laughing man slid over to the seat next to hers. Susan's friend, who knew him, introduced them, and he asked Susan about the restaurant business. She explained that she wasn't really in the business. She was their financial adviser. Her real job was selling insurance and investments.

"Social work for the rich?" It was a good line, and she would remember it.

"You'll never guess what I do," he said.

"You look to me like a Sumo wrestler."

"I have a sex talk show on the radio."

Susan suggested, without mincing words, what he could do with himself. "It's getting late," she told her girlfriend. "I've got to go."

He put out an arm and pulled her back. "I'm sorry. I guess that wasn't the best thing for me to say. But I really do have a sex talk show. Mostly, though, I teach sociology at Boston University."

"Get rid of him. Tell him I'm in a meeting."

"Right," says Norma, the receptionist. She buzzes Susan an hour later. "There's a telegram for you."

Susan goes out to the lobby. Who still sends telegrams? She opens the yellow onionskin paper and runs her eyes over the type: "Looking to merge, forming a new corporation. If interested call this number. Sincerely, Joseph H. Helfgot, Ph.D."

She ignores Norma's inquisitive face. Better to say nothing. Susan has just ended a disastrous relationship with someone in the office, and Norma already knows more than enough about Susan's personal life.

Her phone buzzes again. "There's a package for you. It's from *him*."

Good lord. It's an enormous piñata tied up in clear cellophane with a huge velvet bow. The card says, "Open with care."

Inside the piñata are chocolate-covered fortune cookies with little messages inside. They all say the same thing: *I love you.*

An hour or two later it's a dozen long-stem roses, which she leaves on Norma's desk.

Then: "Susan?"

"Oh, God, what now?"

"Piñata man is on the phone. What should I tell him?"

"Damn it, put him through."

"Hi," he says.

"Stop it."

"No."

"I mean it," she tries to say in a stern voice.

"I thought maybe you would like to have lunch."

"No."

"I'm going to be right around the corner, at the Meridien." He explains that he's having lunch with the president of a major company who is working on her Ph.D. Joseph is her doctoral adviser. Susan knows the company. They have a couple of hundred employees. "Maybe you can join us?"

"No."

"She runs a big shop. Maybe you can sell her some insurance."

"That's low."

April 8, 2009, late afternoon. Intensive Care Unit.
As young Rabbi Franken makes his way into the ICU he spots Jacob Helfgot. The boy's face shows his pain. He has grown taller since his bar mitzvah in September. Driving to the hospital just now the rabbi recalled that festive morning.

Joseph and Susan were with Jacob on the bimah, although both the cantor and the rabbi worried whether Joseph could make it through the full two hours. A nurse stood by just in case, but Joseph got through it. When it came time to address his son, he spoke in a weak but proud voice. Joseph even joked that he and Susan, who had been married by a justice of the peace, had at long last been joined together on a bimah in the presence of a rabbi. Everybody laughed.

In the ICU Susan introduces the rabbi to the family members. He knows about the face transplant; Susan told him yesterday, when he came to visit her. Maybe that's the sensation he's picking up. Or maybe it's just this family, so filled with life as they stand together in death. Even with Joseph's body hovering between worlds, the rabbi can feel his energy rippling through the room, through these people who love him, who are part of him.

They gather in a circle around the bed for the last time, holding on to one another as he recites the prayers for the dying. His voice cracks as he looks down at Joseph's face and envisions a different kind of blessing that this face will soon bring to another man.

Joseph, why are you so warm to my touch? Do you have a fever? Joseph, honey, it's time to get up. We have to go home now. Please, Joseph, please wake up. Emily is crying again. I want her not to cry anymore.

"Ribono shel olam," the rabbi says on behalf of Joseph. "Y'hi ratzon milfanecha she'yih'ye shalom m'nuchati, amen. Master of the universe, may it be Your will that my passing be in peace."

As Susan glances up through the glass, the staff steps back. She sees her family superimposed as reflections on the window.

We are ghosts in the glass, just like you, my love. I cannot bear to leave here without you.

Go away, you people standing there outside the glass. You can't have him!

"Em," she whispers, "we need to go. They're waiting." But still they do not leave.

On the other side of the glass the team tries to remain calm. It has been a long day, and night is about to fall. James Maki has been waiting downstairs since mid-morning. A medical team is en route from another city, and their plane is about to land. Helfgot's heart, already transplanted once in the past week, should soon work its magic again. The moment the family leaves it can all begin. But they will not rush the family, even if they linger all night.

They finally emerge, all but Susan. She stays back, alone with her husband. When she opens the door, her eyes are dry. She stops for a moment to embrace a nurse, then the desk attendant. Dr. Couper is on the phone, and he stands up, trying to get off the call so he can say something. But Susan shakes her head. Not now. He nods in understanding.

As the family turns the corner, medical personnel swarm into Helfgot's room. Every minute counts. Recipients are waiting.

I must go back. I can't leave. I have been coming here every day, for so many days. How can it be that I can't come back anymore? Joseph, how can I leave you here, all alone?

She steps into the elevator.

chapter thirteen

Wednesday, April 8, 2009, 11 p.m. Emergency Room.

finally a lull. Bo Pomahac has been on fast-forward since last night, when Esther Charves called with the news that Helfgot's wife had agreed to a face transplant. He was euphoric, barely able to sleep, but now his elation gives way to a small case of nerves. Each of the thirty-eight members of the medical team knows his or her role, but only Pomahac is looking at the whole picture. If anything goes wrong, and there is so much that can, it will be his problem.

Exhaustion has set in, the kind that hits right after a period of intense anxiety, and the past couple of hours have been *really* intense. Earlier tonight everything came to a halt when a routine presurgical chest X-ray picked up a spot on Maki's lung. It wasn't new. A few months ago, when the same spot showed up on an X-ray, the doctors thought it was just a tiny scar, probably from an infection. But tonight, with so much at stake, the radiology team wants to take a closer look.

Now? Are you kidding me? Pomahac can't believe it.

So Maki was sedated for a bronchoscopy, which meant snaking a tiny lavage instrument down into his lung to brush up some epithelial cells that would be examined immediately for signs of possible malignancy, or any other irregularity. During the past few months he has already been through a workup rivaling that of an astronaut training for space flight: a head-to-toe assessment, including X-rays, CT scans, blood work, the whole nine yards, all of it designed to ensure that he is a suitable candidate for a face transplant.

Then came all the prep work leading up to tonight: the earlier dissections of the cadavers in Brussels, the careful 3-D imaging of Maki's head, and specialized training for the many medical people who will be involved in this enormously complicated surgery. The doctors have established a personalized immunosuppressant regimen to give Maki the best possible chance to fight rejection, and it's a good thing they did. They discovered that their patient was deathly allergic to a common medication given to transplant recipients.

A lot of other people are waiting for this operation to begin, including the recovery teams who are hoping to remove solid organs from Joseph Helfgot's body. A medical team from the Midwest has just arrived to take the newly transplanted heart to a patient who, without it, doesn't have long to live. Professionals from the New England Organ Bank will be arriving later to recover tissue. The film crew is here too. Although their presence is unusual, to say the least, both the doctors and the donor family know that having them in the operating rooms will help make the viewing public more aware of organ donation.

In certain transplant operations, an understudy is waiting in the wings in case the intended recipient is sick or unable to get to the hospital in time. But tonight there is no Plan B, no backup recipient for Helfgot's precious face. This surgery is unique to Maki; the team's practice drills have been tailored to his particular facial defects. Someday, perhaps, Dr. Pomahac may be able to perform a face transplant on a backup candidate, but that day, if it ever comes, is years away. If this surgery doesn't happen to-night, Helfgot's face will be lost. And who knows when Pomahac might be offered another one?

He couldn't believe he might have to go to his patient and tell him they'll have to cancel. Maki heard the good news this morning, and, like his doctor, he has been excited all day. What was Pomahac supposed to tell him? "I'm sorry, Jim, but there's a tiny spot on your lung, and the radiology guys don't like it. It's off for today, but we'll probably get another chance." No way was he going to do that.

They'd had a false start just a few days ago. It looked as if a family who agreed to donate a heart might also be willing to give a face, but they balked when asked about donating back tis-sue. "We're not comfortable with that," they said. If they weren't comfortable with back tissue, they would never donate a face. But on Tuesday night Susan Helfgot had said yes.

Pomahac paced around as he waited for radiology and thoracic to decide what to make of the spot. Dr. Phillip Camp, the director of the hospital's lung transplant program, was still in the building, and when Pomahac had him paged he rushed down to assess the situation. After a few minutes he and the other doctors agreed that the blemish on Jim's lung was nothing but old scar tissue after all.

Pomahac takes a deep breath and lets it out. He needs to lie down. He still has an hour before surgery, which has been pushed back to midnight. But there is no on-call room with a bed on the surgical floor, so he makes his way upstairs to the Burn and Trauma Unit.

"Is there somewhere I can rest for an hour?" he asks a nurse.

"How about Mr. Maki's bed?" They are holding a room where Jim will be brought to recuperate after surgery and where he will likely remain for a few months. The room is immaculate, of course, and the nurse anticipates the surgeon's hesitation. "Go on, take it. They'll change the sheets after you leave." They'll have to clean the entire room, because human skin is full of bacteria and viral agents, not to mention mites, fleas, and microscopic pests that travel through the air and latch onto a person's hair and cuticles. Everyone is a carrier, patients and doctors alike.

Pomahac gets into his patient's bed and closes his eyes, hoping nothing else will go wrong. He can't wait to get started.

Back downstairs Dr. Donald Annino, an oncology and reconstruction specialist who is heading up the surgical prep team, is having a hard time getting into the pre-op area. The transplant is still a secret, and the public relations staff wants to keep it that way until Maki is safely awake in recovery. What if a reporter sneaks in? So Peter Brown has asked to have a security guard at the door. As he looks at the names on his sheet of paper, the guard says, "I'm sorry, Doctor, but you don't seem to be on the list."

"What? Let me see that." Annino grabs the clipboard. "Great," he says, chuckling, as he realizes what has happened. He speed-dials Pomahac's cell phone, jolting him out of his all too brief nap.

"Bo, you won't believe this, but I can't get in. Security doesn't have my name. Not just me, it's all of us. We forgot to put our names on the list."

Pomahac laughs. So much for resting. "I'll be right down," he says.

He hops off the bed and heads out. The nurse, seeing him leave, calls housekeeping to sanitize the room for Mr. Maki. Then she goes in to help strip the bed. She doesn't mind. This is a historic event, and she is pleased to play even a tiny role.

Midnight. New England Organ Bank headquarters.
Ever since Tuesday afternoon, when the doctors signed the death certificate, Joseph Helfgot has no longer been a patient of Brigham and Women's Hospital. Hospital personnel are still deeply involved with his care, but direct medical management of Helfgot's

body now belongs to organ bank employees, who appear on the scene and work with the hospital staff whenever there is a brain death with the possibility of organ or tissue donation. Hospitals are legally required to report all deaths to the nearest organ bank office. Many people register as organ donors when applying for or renewing their driver's license, but sometimes potential donors don't drive, or are too young for a license, or come from another country. In these cases the family must make the decision.

Meredith, the organ bank nurse who took Joseph's oral medical history from Susan yesterday, is watching over Helfgot with the help of the Brigham nurses, monitoring his kidney and liver functions, managing the IV fluids and their composition, and maintaining blood pressure and urine output. In a single moment on Tuesday afternoon everything changed from preserving a life, which was no longer possible, to preserving organs, especially Helfgot's new heart.

From the moment Susan said yes to the face donation, the organ bank has been scrambling to get everything in order. Kristina Andrzejewski, the tissue services manager, sits in semidarkness in her office, trying to grab a little sleep while she waits to hear from the hospital. Soon she and her colleagues should get a call telling them the surgeons are ready for them to come and assist in the recovery of Helfgot's organs, and to make sure they are correctly handled for shipment. There has already been a delay, which is not unusual in these situations.

Andrzejewski has another role as well. Solid organs—lungs, kidneys, livers, and hearts—are removed by doctors, but tissue recovery is done by local organ banks. Arriving with their own scalpels and other specialized tools, trained organ bank employees recover skin, bone, and even heart valves that can be used for reconstructive surgery at some other time, such as skin from the back that is used for grafts on burn victims. When she gets to the hospital Andrzejewski will assess Helfgot's body and determine whether any tissue can be recovered for future use. Unlike organs, which must be transplanted almost immediately, properly preserved tissue can remain viable for up to five years, frozen in a protected vault. If only hearts and kidneys could be preserved this way.

For Andrzejewski it all began with a call late Tuesday night from Chris Curran, the organ bank's lead operations manager. Curran is directing this big and complex symphony, managing every nonmedical detail, including the chartered jet waiting at Signature Aviation in a private area of Boston's Logan Airport. One of his jobs is to make sure that as soon as the traveling heart team returns to the plane with the recovered heart, they will get immediate clearance to take off. This is an urgent matter, as hearts can survive no more than five hours without blood flow.

He is concerned about the weather. It has been a wet and ugly week, and on Monday, Opening Day at Fenway Park was rained out. April is often unpleasant in New England, where climate change can be an hourly event. A plane may be grounded at the last moment, which is another reason they try to line up an alternate recipient in the area, just in case. A healthy, available heart is simply too precious to waste.

Curran is also thinking about the complicated legal issues at play here. The gift of a face still lies outside the framework of the United Network of Organ Sharing, or UNOS, the hub that connects the country's organ banks. Today the whole business has been made even more complicated by the documentary film crew, which adds to the general anxiety that the anonymity of the donor could be compromised.

A few weeks ago the television producer sent a crew to Helfgot's house for an interview about what it's like for a patient and his family to wait for a heart. The crew is aware that Helfgot died the other day, but they have no idea that he is the person donating the face. And the few people who do know want to keep it that way.

With his beard gone and four surgeons and their assisting

nurses working on his heavily draped body, not even his wife would be able to recognize him. But the hospital and the organ bank want to be extremely cautious, so the organ bank's lawyer has prepared another set of forms that Susan Helfgot signed on Tuesday, agreeing not to hold them, the television network, or any caregiver or institution responsible if her family's anonymity is compromised.

Curran managed to make it to tonight's Red Sox game in the freezing rain, returning home just as Pomahac was starting his abbreviated nap in Maki's bed. He is glad to get out of the cold and back to his computer and his landline. He looks at his cell phone, half-expecting it will ring, but hoping it won't. Sometimes a donor becomes hemodynamically unstable. If Helfgot starts to crash, Curran will be involved in some quick decisions about which organs can be salvaged, and in what order.

Midnight. Hospital basement.
Dr. Julian Pribaz makes his way along the cold corridor. Midnight may seem like a strange time to start a cutting-edge face transplant, but he doesn't mind. He enjoys nights in the hospital, when he can think his own thoughts without the constant din to distract him. The elevators are empty, the hallways are silent. There is no hint that tonight will be unusual.

Pomahac called him last night. The two surgeons are as ready as they'll ever be, and their donor is right here at the Brigham.

Having Helfgot in the house is a big advantage; there will be no need to orchestrate events between hospitals that could be hundreds of miles apart. This operation will be complicated enough, with other organs being recovered at the same time. The out-of-town heart recovery team is already here for a surgery that was supposed to start at six o'clock. Now it's almost midnight. They can't be pleased about the delay.

Although he and Pomahac practiced in the Brussels lab, with

Pomahac on the right side of the face while Pribaz tackled the left, they have never handled a real face donor whose blood still moves throughout his body. Pribaz relishes this final precious moment of contemplation. As he enters the locker room, where he will change into fresh surgical scrubs, he is overcome with anticipation. He slips off his lab coat, hooks it inside the narrow locker, and slams the thin metal door.

Upstairs Dr. Pomahac enters the operating room just before Helfgot, warm and pink, is wheeled in. The only sound comes from the respirator, which is pumping air in and out of his lungs. Other than Pomahac, there are only two people here: the ventilation specialist and the gurney operator, who is responsible for the critical job of moving a body that is being kept alive by artificial means. This is no simple task, as it involves protecting all the equipment that is being moved along with Helfgot: the respirator, the vital-sign machine, drainage bags for waste, and IV lines. Even the gurney is not really a gurney, but an adjustable bed from the ICU.

Pomahac begins to wonder. Where the hell is everybody? Has the surgery been called off while he was napping? Is it that damn spot on Maki's lung? In a mild panic he calls Julian Pribaz.

"Where are you?"

"In the locker room."

"Oh, good."

"Don't worry. I'll be right there."

Thursday, April 9, 2009, 2:00 a.m. Helfgot residence.
I can't get myself to lie down on this bed. It's a dead man's bed. So I sit on the edge and pick up the picture we keep on the night table. We're together in Laguna Beach, near the water, with another couple. We

look pretty good, the four of us, darkly tanned with healthy smiles in the brilliant California sun. When was that, ten years ago? It feels like a lifetime.

I hear Ben rushing up the stairs. His bedroom door slams closed. He's talking loudly on his cell phone. It feels like daytime, but the clock tells me otherwise. We all know what's going on at the hospital. Who can sleep at a time like this?

Wonderful Ben. He'll leave in a few months for New York. I still can't believe he got into NYU. Joseph was so happy. Thank you, God, for letting him be alive on that wonderful day. Long before we met, Joseph used to teach there. Once we sat for hours in Washington Square Park, across from where he taught sociology, and he told me about those early days. It was freezing, but we didn't care. We had just met. You don't feel the cold when you're falling in love.

So he won't be here to see Ben go off to college. I can't believe Jacob is starting high school, but at least he'll still be home with me. Thank you, God, for not taking all of them away from me at once. I just couldn't bear it.

Please make this night be over. My heart is racing as I keep thinking about Joseph lying on an operating table in a cold room, being pulled this way and that. I don't even know these surgeons, and Couper's not there. What am I thinking? Couper doesn't need to be there. Joseph's dead. They can't hurt him.

But his face—I see it happening. It's too much. I get dizzy and sink down into Joseph's pillow. It smells just like him. Maybe I should have thought about this some more. Pam told me to slow down. But I was so damn sure it was the right thing to do. How can you not give somebody a chance to live a normal life? Dr. Lewis was so excited when he called us about the heart. In the blink of an eye, everything was going to be all right again. And then the other eye blinked.

They've turned on the TV in the kitchen. What is that theme song? Maybe I should go down there and be with them. But I just can't move. Will they call me from the hospital when it's over? Esther promised they *would. It's two in the morning. They should be almost done by now. God, please make this go quickly.*

It helps to think about this poor man, the recipient. How can you lose a whole face? Esther said it was a horrible accident, that he had been a housepainter. Maybe he fell off scaffolding? She said he can't eat or breathe or talk. Will this surgery save him? I look again: it's 2:10. I'm not going to sleep. There's no way. I close my eyes and see a scalpel tracing Joseph's face. I'm so cold I'm shaking. I pull the covers up over my clothes.

chapter fourteen

Thursday, April 9, 2009, 1:15 a.m.
Helfgot Operating Room.

he first trace is made with a fine scalpel that outlines a portion of Joseph Helfgot's face. In Brussels it took them just over an hour to recover a face when they practiced on their fourth and final cadaver. But tonight will take much longer because they're working with a heart-beating donor with blood flowing through his body. A body undergoes less blood loss after its heart has stopped beating, but that's not true for donors with lifesaving organs. Blood loss must be minimized to prevent Helfgot's blood pressure from crashing, which would compromise his solid organs. If they lose the heart, it would be a disaster.

It doesn't help that the human face is heavily vascularized, with large carotid arteries feeding smaller and smaller vessels, the arterioles and capillaries that take blood to specific locations, each accompanied by counterpart veins that carry it back to the heart. These passageways serve different areas of the face, each with its own sets of muscles that go to work when your brain tells you to chew or swallow, to smile at a baby or wink at a friend, to stick out your tongue or grimace in pain. The phone rings with a friend on the line, and you talk and laugh or cry, holding the phone while sniffing and then tasting the spaghetti sauce on the stove with no thought about the millions of brain signals and controlled muscle twitches that make it all happen. These small, simple, ordinary motions of a face require large amounts of energy gathered from lots of richly oxygenated blood. Ask anyone who has ever had a nosebleed.

The luscious beard that Helfgot had sported since he was a teenager was shaved a few hours ago. It's already trying to grow back, because hair follicles don't need a brain to tell them to go to work. His transplanted heart can also function without his brain: it has a tiny cluster of nerves centered high in the right atrial chamber that will beat, at least for a while, in a last-ditch but hopeless attempt to prolong life. As long as the ventilator supplies the lungs with oxygen that can attach to hemoglobin and travel the bloodstream, keeping the cells healthy, the heart will continue to beat. The human heart is a miraculous instrument, tough enough to fly off to another city and take up residence in the chest of a third person, and keep him going.

Pomahac's team will do everything they can to protect this heart, keeping it strong and healthy during the procedure. Jim Maki probably won't die if he has to wait six months or a year before another donor family offers the gift of a face, or even if a face never arrives. But the man waiting hundreds of miles away may have only a few weeks to live. If a new face means almost everything, a new heart means *everything*. If Helfgot becomes unstable, it's an easy call: the heart trumps the face.

Pomahac knows what's at stake as he works with Julian Pribaz, Elof Eriksson, and other members of the recovery team as they make the first incision. The key is to finish with minimal blood loss. If they fail, and the heart is damaged—well, they all know what that means.

Maybe they should do this in stages. Maybe, Pomahac is thinking, they should stop halfway, after they raise the flap, a section of skin that includes the veins and arteries and carries its own blood supply. They could pause and let the other teams take Helfgot's solid organs before they go deeper into his head to remove the bone that will restore structure to Maki's face. If there is excessive bleeding during surgery it's likely to happen then, when the bone is being cut, which is why it's not unusual for a patient to receive a blood transfusion right before major orthopedic surgery. The transfusion acts as a reserve in the event of major blood loss during surgery and protects the blood pressure from a precipitous drop.

Either way, they will need to remove a large portion of Helfgot's facial bone, cutting into the zygomas on each side near the maxilla intersections. They also need his hard palate and upper teeth, which will require cutting into the pterygomaxillary junctions, then all the way up to the top, through the orbital floor of bone that encases and protects the eye. Maki lost a lot of bone, which was burned off when he was electrocuted, leaving a deep, dark void in the center of his face.

In spite of the massive amount of bone work required, they finally decide to do it all without stopping, unless, of course, the heart becomes threatened. Using a syringe they squirt epinephrine into Helfgot's cheeks to constrict blood flow while they move their scalpels deeper into the face, isolating the nerve bundles they'll need along the way. Then they'll move on to the muscles, watching out for preferred arteries and veins before cutting the mandible bone to expose and recover the facial arteries.

They work through the night, but Pomahac is not tired, not at all. He and his team are operating on pure adrenaline.

• • •

7 a.m. Helfgot Operating Room.

Kristina Andrzejewski from the organ bank enters the OR with two coworkers. Everyone assumed that somebody must have called her during the night with news of another delay, but as dawn broke, nobody had. And so the three of them drove to the hospital to start reviewing Helfgot's chart, a heavy binder with reports of everything that happens to a patient during a hospital stay.

Helfgot's heart doctors are now busy with their other patients, and organ bank staffers assume nothing. Andrzejewski will pore over the chart and make an independent assessment as to whether or not Helfgot's tissue is healthy enough to be preserved.

Because nobody involved in removing Helfgot's face knows anything about his medical history, his chart is essential. Andrzejewski reads through it carefully, looking for red flags that will prevent a potential tissue donation, such as a history of hepatitis C or a latent infection. A recipient can be given antibiotics to fight an active infection, especially when a donated organ spells the difference between life and death. But valves, bone, and tissue that will be preserved for later use are in a different category; they cannot be given antibiotics in a preserved state. Most organic materials are considered medical devices by the FDA and are held to an extremely high standard. If there is any sign of latent infection the material will not be taken.

It takes Andrzejewski more than an hour and a half to go through the three four-inch binders that make up Helfgot's chart. She has read a lot of charts over the years, but she has never seen one this large. She can't believe what this poor man had to endure over the past eighteen months. In the end a previous infection will rule him out as a tissue donor. But she and her team are required by law to stay and observe the operation, ensuring that the heart, kidneys, liver, and pancreas—if they're healthy enough—will all be properly recovered and packaged for delivery. The face too, even though it's only traveling across the hall.

When Andrzejewski enters the operating room she makes eye contact with Dr. Pomahac. They first met two years ago, when he visited the organ bank to ask for their help in expanding the range of organs to include the face.

A monitor on the wall shows the surgery in progress on the table. It's hard to make out what's really happening, but she sees that Pomahac is flipping Helfgot's facial flap back and forth like a pancake, looking at it from the top and then viewing it from the bottom. The flap is still attached to Helfgot's former face by a small pedicle of vessel-rich flesh. Pomahac checks everything one last time until he is satisfied that it's safe to detach.

Andrzejewski watches her staff. Going in, it is impossible to know how they will react to a face recovery. She can handle it, and she's pretty sure they can too, but it's new territory for all of them. She and her associates are ready to document the first New England Organ Bank protocol for packaging a face. There will likely be more faces coming along in the future, and they will almost certainly need to travel more than a few feet.

Andrzejewski has one last task, and this one is personal. She has promised Esther Charves that she won't leave Helfgot's bedside until he is safely on his way to the funeral home. She is now his protector.

Regina Swanton, who works with Andrzejewski, is tense. The documentary crew have their faces in their cameras, just inches from the operating table and workbenches. She worries that they might accidentally brush against a sterile field, although she can't blame them for trying to get the best possible view. She wonders if anyone else is concerned, but if so, there's no hint of it. These videographers have been filming in operating rooms for months, and they know the drill.

At 7:02 in the morning they are finally done. It has taken almost six hours to detach the parts of Joseph Helfgot's face that will give James Maki a new life. What was a face when they started is now officially known as an allograft. Temporarily in limbo, it awaits its new owner, who will mold it into his own unique identity.

7:30 a.m. Maki Operating Room.
Pomahac needs to check a few things, especially Maki's arteries and veins, which must match up exactly. Maki is completely anesthetized, and for several hours another surgical team has been preparing his face for the transplant, which is just minutes away. Although they have been on their feet and under the operating lights for hours, in a sense their work is only beginning. The next stage will be far more strenuous, because unlike Helfgot, Maki is alive and must remain stable on the operating table for another eight to ten hours, maybe longer.

Pomahac looks down at his patient. He knows every square inch of this face, which bears the terrible scars of the ten surgeries he has performed on it over the past four years. Until tonight a large flap covered the section where Maki's nose and mouth used to be; Pomahac attached it during a previous surgery to cover the gaping hole created by the accident. Now, finally, it has been removed and discarded. But there is a lot of scar tissue, and cutting through heavy scars is never easy. Scars are nature's way of holding damaged skin together, and they will make for slow going.

The allograft is in a tray on a table in Helfgot's operating room, its nerves tagged and blood vessels purged of possible clots before being readied for transport. The doctors must get Maki's blood flowing through it as soon as possible, before the cells begin to die. The aorta brings blood up to the head in swift, strong pulses, through two sets of large carotid arteries located on both sides of the neck. Each set has two branches, the protected internal carotid feeding the back of the head and the brain, and the external one, which feeds the face.

They have decided to use Maki's left external carotid as the main source of blood for the allograft, matching it up with Helfgot's left carotid that was taken along with the allograft tissue. The two sections will be sutured together in a microsurgical technique known as anastomosis. If all goes well, blood should begin to flow from Maki's carotid into the allograft, perfusing it back to life.

If the left carotid fails, they will need an alternative blood supply. Maki's external right carotid is not an option, as it is busy feeding blood to the rest of his own face during surgery. Once they start dissecting his left carotid artery and preparing it for anastomosis, it will be almost impossible to turn back.

Now that his colleagues have exposed it, Pomahac examines Maki's artery to judge its lumen, or circumference. It isn't wide enough to suit him, so he instructs the team to go a little farther along on the carotid to enlarge the vessel cavity. That should help the blood flow more easily.

If we can get this first artery to hold, he is thinking, *we can start to relax.* The biggest danger is a blood clot. There is always the risk of a clot, and clots kill. A few of them snaked their way into Helfgot's aorta while he was receiving his new heart. They traveled through his carotids and into his remarkable brain, shutting down its blood supply and ending his life. In medicine, anything is possible.

7:40 a.m. Helfgot Operating Room.
Before Pomahac returns to Helfgot's operating room, he and Julian Pribaz confer. The brief break allows them to catch their breath before pressing on. Pomahac looks up at the clock. It's close to eight, and there is still a whole surgery left to perform.

He returns to help Andrzejewski package up the face for transport as carefully as if it were going to another city, rather than merely across the hall. The allograft has been bathed in Wisconsin Conservation Fluid, better known as UW solution. This special liquid, developed by two Wisconsin doctors in 1987, matches human pH almost exactly, with a rich mix of chemicals that will keep the cells stable until Maki's blood flows in and takes over.

Pomahac wraps the allograft in sterile plastic and then wraps it again in more plastic. With a gentle motion he eases it into a tray filled with a slurry of ice. Then the package is wrapped in plastic yet again, twice, to ensure that no virus or bacteria can attach itself to the human substrate during its brief journey across the hall. Andrzejewski watches carefully to guard against any mishap or confusion. There have been one or two errors in the sixty years of transplant surgery in other parts of the world, with consequences similar to those of a hospital discharging a baby to the wrong parents. But nobody is really worried that New England's first facial allograft will wind up on the wrong person.

A blue plastic tote container that could just as easily hold a few beers serves as a temporary sarcophagus for the face. Pomahac snaps it shut and carries it out of the room, trailed by the camera crew and a procession of surgeons and residents. He enjoys the feeling of knowing how well protected the allograft is. It will be safe even if he slips.

While Pomahac has busied himself with arteries, veins, and UW solution, another team of surgeons has been busy under the direction of Dr. Christian Sampson. They are removing skin from Helfgot's forearm to be used as Maki's sentinel flap, which is a small piece of skin from the donor that is grafted onto the recipient in a discreet location. (In the French transplant Isabelle Dinoire's sentinel flap was placed under one of her breasts.) At regular interviews after the surgery, doctors will snip off small pieces of the flap and examine them.

These tiny biopsies are studied for signs of possible rejection in the recipient in the form of immune cells that show up and function as early storm clouds, appearing well before the rest of the body reacts to the rejection alarm. If necessary the immunology team can launch a preemptive attack with large quantities of drugs, such as prednisone, that will keep the immune system from rejecting the new organ. Creating a sentinel flap protects the transplanted area from frequent biopsies. Maki's face has already been through far too many assaults, but this one is avoidable. A sentinel flap also provides a second indicator: if it appears that rejection is forming at the site of the transplant, but the flap remains unaffected, it may be just an infection.

For Maki's sentinel flap, Dr. Pribaz has come up with an inventive solution. Maki had severely damaged his right hand when he fell onto the tracks, and his thumb and index finger were badly burned and fused together. Pribaz wants to use a piece of Helfgot's forearm skin to open up the area between Maki's thumb and index finger. This webbing can also serve as the sentinel flap. Sampson's team has been working on the flap, and it too is now being prepared for transport across the hall. On any other day the new hand flap would constitute a significant medical procedure. Today it seems like an afterthought.

9:15 a.m. Maki Operating Room.
Dr. Pomahac brings in the face. The others in the room are excited, but their expressions lie hidden behind their masks. A pity, thinks the videographer, who is trained to pick up the slightest nuance—a furrowed brow, a wince, a bead of sweat. Even the surgeons' eyes are hidden behind large optical lens machines that hang from the ceiling, which can magnify the tiniest blood vessel or nerve. So much is lost when you can't see someone's face.

The surgery starts in earnest, with Pomahac performing the

delicate anastomosis that connects Maki's left carotid artery to its allograft counterpart. More than two dozen people are trying to crowd around the table, but there isn't room for them all. Finally the microscopic sutures are in place. Two years of work have gone into this moment: cadaver trials, protocol issues, recipient approval, and the long wait for the right donor.

Pribaz looks up at Pomahac before he slowly releases the clamp holding Maki's blood supply at bay. This is it, his eyes are saying. The room falls silent. Everything has come to a complete stop as they all watch. And they wait.

Finally, almost imperceptibly, starting on the left side of the cheek where Pomahac has attached the vessel, the pale white allograft begins to turn pink. Gradually color creeps across his patient's new skin, slowing down a little as it struggles against gravity at the bump of the nose, then moving faster down the other side, toward the right cheek. A pink tinge blossoms on a waxen field.

It's as if a spaceship has just landed on the moon. But nobody cheers. Nobody says a word. They are watching a dead man's face being resurrected, which they recognize as a kind of miracle, a Lazarus moment. But there are no shouts of joy, no smacking of backs. The medical team is humbled, and some of them blink back tears. But only for a moment, because there is plenty left to do.

There are more blood vessels to attach—arteries to pump blood and veins to take it away. The surgeons join Helfgot's right jugular to Maki's major facial vein and continue until their work is done, until the allograft is finally vascularized and blood flows in and out of what is now Jim Maki's new face.

The hand surgeons have started attaching the sentinel flap. While they work, Pomahac and Pribaz begin to connect nerves on either side of Maki's face, using a technique known as neurorrhaphy. It is painfully slow going as the doctors knit together

Helfgot's and Maki's tiny nerves, one suture at a time. Nerves near the surface will provide sensation, allowing Maki to feel steam rising from a cup of hot coffee or a light breeze on a warm day. Other nerves, sutured deeper in the face, will one day allow him to chew and swallow. They are piecing together a kind of 3-D jigsaw puzzle, one tiny segment at a time.

Minutes ago, when the allograft first perfused, forcing blood through Maki's facial vessels, the adrenaline pump in Pomahac's body finally began to relax. Now that he can afford it, deep exhaustion starts creeping in. His last normal sleep was on Monday, three nights ago. He leaves the room and scrubs out for a while, allowing the hand team and everyone else to continue without him. He and Pribaz have picked the whole team, and they trust them completely. What he really needs now is a cold Diet Coke, or whatever is still left in the vending machine.

Pomahac's team starts working on the upper palate and teeth, and the rest of the bone. It is attached with titanium plates less than two millimeters wide that will give structure and protection to Maki's underlying facial area. It will allow him to have a mouth again, so he can eat. And smile. And speak.

11:00 a.m. Helfgot Operating Room.

The traveling heart team has been awfully patient. True, they've had no choice, but they've been waiting here since 11:15 last night. They can't sleep, because if Helfgot becomes unstable they'll need to move quickly. For the past hour they have watched the hand surgeons removing a piece of Helfgot's arm for the sentinel flap. Now, at long last, the room belongs to them.

They are feeling some anxiety, which is normal at such moments. Before they flew to Boston Dr. Couper told them that the donor's heart appeared to be sound, but it's impossible to know for sure without actually seeing it. They hope Couper is right.

Back at their own hospital the intended recipient is hoping the same thing.

Helfgot's body is still split open from his heart transplant three days ago. A long, four-inch-wide gash runs down the center of his chest, covered only by a fine-mesh elastomer membrane, a plastic that acts and feels like rubber. Heart recipients are given so much fluid during surgery that it is often impossible to close up the chest right away because their bodies are so swollen. Leaving the body open for a while also makes it easier to go back in if there's a problem.

Sometimes patients are left open because of all the hardware. When Helfgot died he was full of metal: a VAD, a defibrillator-pacemaker unit, and an experimental device for monitoring heart fluid. Only the VAD was removed during his transplant. Couper had intended to remove the rest later.

Working with the heart team, two Brigham surgeons isolate Helfgot's organs by dissection, separating them from their blood supply and flushing them with fluid. The first to be recovered are the liver and pancreas, but a quick biopsy of the liver shows it to be fibrotic. Someone calls the organ bank to tell them the news. The potential recipient will have to hang on a little longer, hoping that another liver will come in time. But Helfgot's liver is not wasted. A staffer at the organ bank calls around and finds a research team that is happy to take it.

The pancreas has been damaged from years of strong medications that led to diabetes, but it too finds a home in a research lab. As Esther suspected, Helfgot's kidneys are compromised from years of diuretics to purge excess fluid buildup from his sorry heart. Helfgot was sixty, but his kidneys look like those of a man of ninety. His corneas are not taken because of early-stage cataracts. And because he was exposed to a recent drug-resistant bacterial infection there can be no tissue donation.

But that still leaves the most important organ of all. At 11:25

a.m. Helfgot's major aorta is cross-clamped. His heart finally stops beating. One of the heart recovery surgeons scribbles the time and day on the pant leg of his scrubs. The five-hour countdown to transplant has begun.

After a quick visual check, the heart is carefully pulled and immediately placed into solution to protect the cells. The ambulance on alert outside the hospital starts up. After a call from the organ bank, the pilot of the small Cessna secures his final flight plan, preparing to depart within half an hour. They have five short hours to get from the Brigham through the Boston Harbor Tunnel to Logan Airport, board their jet, fly to their home city, and race by ambulance back to their hospital. A quick goodbye and they're gone.

It's all over. Everything that can be removed is gone. With a flip of a switch, Helfgot's respirator is turned off. It is now up to other specialists to restore some small semblance of dignity to this man who, until two days ago, was one of the hospital's favorite patients. The body is closed.

chapter fifteen

*Y*ou fell in love in a bathtub?"

"Joseph was living in Cambridge, right in Harvard Square. He had this tiny, postage-stamp condo with an enormous cast-iron bathtub, where the water stayed hot for hours. You know, the kind with claw feet?"

Susan is on the back porch with two of her girlfriends, who have dragged her away from the insanity inside the house. A man is stacking cases of wine against the wall. There's a beeping sound in the driveway from a truck backing up. The women look up as two men come around the back of the house. One is carrying a round tabletop and the other has several folding chairs hooked under his arms.

Who ordered furniture? We're not having a wedding. And why do we need all that wine? There's still some in the basement from Jacob's bar mitzvah.

"In a bathtub?" asks one of the women. "It must have been a hell of a tub."

"No, it wasn't like that. I went to his apartment to meet him for dinner. We were going to walk around Harvard Square and get something to eat, but it started raining like crazy. So he picked up takeout at the Chinese place next door."

Someone has uncorked a bottle of white wine and poured it into plastic cups. Susan takes a sip. "Ugh, this is awful, it's warm," she says.

"Drink it. So you're in the tub . . ."

"So we're watching a movie and eating takeout. I wish I could remember what movie it was. And then we started making out."

"*Finally,* the good part."

Susan shakes her head. "Not really. My lips started blowing up. And then I couldn't swallow. He had ordered soft-shell crab, and I'm allergic to seafood. Of course I didn't have any, but with all that kissing . . ."

"You had a reaction just from kissing?"

"There was a *lot* of kissing." Everyone laughs.

"Did you have to go to the hospital?"

"I came right from work, and I had my briefcase with me, with my EpiPen. Joseph ran and got it and I gave myself a shot. In a few minutes I was fine."

"Jesus."

"It's a good thing I had my briefcase."

"And *then* the bathtub?"

"I had tiny hives all over, like poison ivy, but worse. I couldn't stand the itching. Joseph was so freaked out. He was flitting around like a bee, completely frantic. It was so sweet." She smiles at the memory.

"He ran to the pharmacy to get me some cream and an oatmeal bath. While he was gone I filled the tub and got in.

He came back with four bags full of stuff. He must have bought one of everything. You know Joseph." They all nod. "He just stood there with those bags, looking at me. His eyes were all red. He had been crying. I didn't know yet that he cried so easily."

"Oh, Sue."

"He said, 'Would you like me to read to you while you sit in the tub?' I just remember looking at him and our eyes locked. That was it. We just knew."

The women sit quietly, the sound of occasional laughter mingling with clinking china coming from the house. The air is heavy with spring. The rain has finally moved on and the sun is fighting its way out of the haze.

Someone pops her head through the screen door. "Sue, telephone. I think it's the funeral home."

"Mrs. Helfgot? We're trying to get an update on your husband. Do you know when he's supposed to get here?"

Susan glances at the clock on the oven door. It's almost four. "He's not there yet? I don't know. Maybe you should call the organ bank."

"Don't worry, Mrs. Helfgot. We'll figure it out. Sorry to have bothered you."

Susan hangs up the phone. The surgery started last night.

They're not done yet? Where the hell is my husband?

Thursday, April 9, 2009, 1:30 p.m. Helfgot Operating Room.
Kristina Andrzejewski is still standing watch. She has been asked to notify Dr. Suzuki when they were ready, and he has arrived with the mask. He examines the space where Helfgot's face used to be, looking with concern and disbelief because the cavity is much bigger than he expected.

Andrzejewski watches him remove the silicone mask from its

container and begin to fit it on the ravaged hole that is now Helfgot's face.

"Much too small," he says. There is a huge gap on the right lower chin of almost five inches.

"I have to fix this. This is not good."

To cut the tension, Andrzejewski asks, "So what's your specialty?"

"I'm a dentist."

A *dentist?* Her surprised look forces an explanation.

"I'm Mr. Maki's prosthodontist. I'm a professor at Tufts Dental School."

"Oh." Because she has never met a prosthodontist in her line of work, it takes her a moment to remember that they specialize in implants, jaw problems, and restorative work. Who better to make a mask, she decides, for a person without a face?

She leaves Dr. Suzuki to his work, but she is fascinated by what he's doing. The mask is creamy white, very *Phantom of the Opera.* It's amazing how real it looks as Suzuki matches it to Helfgot's cheeks.

"It's not right," he says. "Bo said he had to take more than he expected, but I had no idea."

He takes out some liquid silicone and begins working on the side of the right cheek, trying to fill in the space. But it's just too large. He pulls gauze from his bag and packs the area underneath, stuffing it into the cavity so the silicone won't flow back into Helfgot's head when he pours it into the gap.

Silicone takes six hours to dry completely, but he doesn't have that kind of time. Helfgot's body has to arrive at the funeral home before sundown, when the second night of Passover begins. The funeral is tomorrow morning. He watches as clumps of wet silicone fall slowly back into the gauze. Silicone oozes from the sides of the mask and sags down the neck. Frustrated, Suzuki wipes some away and starts to apply fresh silicone.

For the next few hours he repeats the process again and again, becoming slightly panicked as he realizes he is running out of time. It's almost four o'clock.

Andrzejewski thinks the mask looks fine, but Suzuki is a perfectionist. A nurse is working to get the room ready for the next patient. This operating room has been tied up for sixteen hours, and so has Maki's. The faster she can restock it and sterilize it, the sooner they can start to ease the backlog.

With two operating rooms and eight surgeons dedicated to the face transplant for more than a day, several other surgeries have been canceled. A few out-of-town patients arrived this morning only to learn that their procedures will have to be rescheduled. Later there will be phone calls and letters of apology, but no explanation. Although these patients might feel better if they knew all the facts, federal privacy laws forbid any such disclosure.

Suzuki continues to add more silicone. "This is not good," he mutters. He is becoming more and more upset.

The nurse tries to encourage him. "I never thought I could donate my husband's face, but what you're doing makes me reconsider," she says. It's hard to know if he hears her. He shakes his head, his brow now beaded with sweat. "It's no good," he says again. He is angry now. "I promised them I would make it perfect."

"It's really there to dress the wound and show respect," says Andrzejewski. "You've done the best you can. We have to get Mr. Helfgot there before sundown."

It is now 4:45 p.m. Suzuki has been hard at work for three hours, trying to make the mask fit.

"That's it," he finally says. "There's nothing more I can do."

He takes out his paints and begins to work on the color of the mask. Pomahac had told him to make it light, but he is unprepared for how pale the actual skin surrounding the mask has become. He needs to lighten it up, but his paints aren't up to the task. In the end the mask is a bit pinker than the pallor of Helfgot's forehead and neck.

The silicone has still not set, so he takes more gauze and wraps it around the head and face. Helfgot looks a bit like a mummy. It's an unsettling image, but it will have to do.

The hearse has arrived to deliver his patient to the funeral home. *His* patient, because over the past three hours Suzuki has become possessive. He worries about what will happen on the ride to the funeral home. The new part of the mask is like jelly; it still hasn't set. If it is jostled, all his careful work will be destroyed.

But it's time to go. He gently cradles the head and helps lift Joseph Helfgot off the table and onto the gurney for his final ride out of the hospital.

Suzuki leaves exhausted and defeated, although he has been told that the family does not plan to view the body. As he drives home he makes two vows. The next time, if there ever is a next time, he will make a full facial mask and trim it down. And to help set the silicone he will bring a blow dryer.

Kristina Andrzejewski notes the time as she calls her office to say they are finished. It is five o'clock.

Across the hall. Maki Operating Room.
The allograft skin is perfectly tailored. The surgeons have left extra room for the inevitable swelling caused by the tremendous assault to Maki's face during the surgery and from the antirejection meds he will have to take. Pomahac will tighten up the face in a future surgery. But for now, the fusion of two faces is complete.

Around the time the last suture is placed, a hearse pulls into the funeral home with the body of Joseph Helfgot.

chapter sixteen

he team has been guardedly optimistic since finishing up the surgery. It took a long time and seemed to go almost perfectly. Jim Maki passed the first hurdle when he was roused this morning. He was groggy and slightly disoriented, but no worse for wear after seventeen hours under general anesthesia.

The blood flow to the allograft has continued, and so far it is perfusing well. No clots or blown vessels. Maki's nurse, Lorrie MacDonald, is watching him like a hawk for any possible signs of trouble. His vital signs are good. He is remarkably fit for a man of almost sixty with a history of heroin use.

If there are problems after the surgery they are likely to appear during the first twenty-four hours. But one big worry must remain unaddressed for at least a few days. Will Maki's immune system reject his new face? It will certainly try, but to what extent will it succeed? How much havoc will all those cells coded with Joseph Helfgot's DNA wreak on James Maki?

The doctors have matched donor and recipient by age, sex,

and blood type. Their tissue matches well too, sharing several antigens that should help reduce the risk of acute rejection. On paper the two men share a lot. But Maki is half-Japanese, mixed with what appears to be Native American and Caucasian heritage, while Helfgot was a Polish Jew. They come from completely different gene pools, and Maki's immune system will soon be aware of it.

Immunity is a complicated business. Millions of complex chemical signals cascade in perfect sequence as cells combine in a variety of patterns in a frenzied dance to keep invaders at bay. In addition to killing off harmful bacteria and viral agents, the human immune system will also attack foreign tissue and organs. It doesn't know that a new heart or a new face has entered its country legally, that its passport is in order, because until very recently in the course of human history nothing like this has ever happened. Without asking questions, the immune system acts swiftly and without mercy, mounting a full-scale assault to destroy the presumed invader.

There are several branches within the immunity army. T-cells, some of which are known as natural killers, punch holes in virally infected human cells. B-cells morph into antibody-producing machines that tag invading bacteria for destruction by other cells, known as macrophages. Dendritic cells with octopus-like tentacles announce an invasion much like the family cat presenting a dead mouse on the doorstep: they curl up foreign protein into one of their long arms and bring it to a T-cell, marking it for destruction. Along the way these various cells combine with potent chemical cocktails. Some of them seem to remember enemies from previous battles, while other recruits serve as scouts in the hunt for new intruders, like the H1N1 virus or, in Maki's case, a transplanted nose.

The human immune army is on active duty every second of every minute of every day. It knows to ignore visitors with special diplomatic status, like the bacteria in the human gut that aid in digestion. It will detect and attempt to destroy previously healthy cells that have turned renegade and become precancerous. Rejection can occur in an instant. Nobody has yet figured out how to either persuade or trick the body into believing that transplanted organs and tissues are on the guest list.

Someday scientists hope to replace faulty organs with substitutes that are grown from a person's own stem cells. They have managed to grow ears on the backs of mice, but these ears cannot hear. They can grow a human bladder from bovine cells and are learning how to integrate it into the rest of the urinary tract. For now, donated human organs are the only viable option, and rejection is minimized with powerful drugs that slow down the immune system. But there is a steep price to pay. These same drugs suppress the body's response to infection, which compromises patients for the rest of their lives.

Many transplanted patients who would otherwise die live on for decades, but they face higher rates of diabetes, kidney disease, and cancer. Even with these risks James Maki was eager to receive a new face. Dr. Pomahac has several other patients whose injuries are as severe as Maki's, and they too are hoping for new faces. Assuming that Maki continues to do well, they may get the chance to undergo this life-giving surgery. All they will need is a brave donor family, a fair-minded insurance company, and the continued blessing of the hospital's Institutional Review Board.

Because Maki's surgery is new and mostly untried it is considered a research trial and is subject to IRB scrutiny. This is standard operating procedure; every facility in the United States that conducts medical research on human beings must follow strict guidelines established by its own IRB. An independent panel determines what type of research will be permitted within a given institution and defines its parameters. No two IRBs are identical. Dr. Maria Siemionow, who performed the face transplant at

Cleveland Clinic four months ago, has her own set of guidelines, as does Bo Pomahac at the Brigham. Often there are different panels with particular expertise, and the Brigham shares several IRBs with Massachusetts General Hospital, across town. Each of these panels is made up of twenty to thirty members, including not only medical specialists but lawyers, ethicists, statisticians, academicians, business leaders, and other concerned community members.

The goal of these panels is to maximize potential benefits while minimizing potential risks, but achieving that balance is never easy. Originally Pomahac approached the Brigham's IRB panel with the idea of limiting face transplants to patients already on immunosuppressant medication. In other words, only people who had previously undergone a transplant would be eligible. Although this wasn't a practical solution because the population involved was so tiny, it was a good way to begin the conversation.

By 2008 it was clear that Isabelle Dinoire, the French recipient, was thriving almost three years after her surgery. Her outcome was spectacular, affording her a quality of life that was impossible to achieve by other means. She appeared to have suffered no adverse physical or psychological effects. In view of her progress, Pomahac revised his application to offer a partial face transplant to anyone with a medical need, without regard to ongoing immunosuppressant drug therapy. The IRB met again, and Pomahac prepared for the questions he thought they might ask:

Do you have a particular candidate in mind?

What will you do if, unexpectedly, the recipient ends up looking like the donor?

How can you know if someone is psychologically ready for a face transplant?

There would surely be medical questions about rejection therapy, questions about the surgery itself, and perhaps even one or two about protecting the identity of the donor family. No IRB question is inappropriate, but there could be a response by the principal investigator—Pomahac, in this case—that would send him back to the drawing board.

In the end, after much discussion, Pomahac's research trial, opening up eligibility to people like James Maki, was approved by a majority vote.

Now that the surgery is over the IRB will watch and wait. They will monitor Maki's progress and will act as his guardian to make sure he is protected going forward. So far they have given Pomahac permission to perform only this one face transplant.

Pomahac comes in to check on Maki, who is still struggling to clear his head from anesthesia. The surgeon can see something the IRB panel doesn't yet know: the results look promising. He is certain that the man who has endured the horrified stares of children, and who has gone without a real meal for four years, will be pleased.

Pomahac looks at his watch. The press conference is just a few minutes away, and he'll have to stand up in front of all those people. Peter Brown is looking for him; he is relieved that Peter has his back. As he leaves Maki's room he tries to compose himself into a subdued look, to repress the ear-to-ear grin that is trying to take over his face.

Office of Public Affairs, Brigham and Women's Hospital.
Late yesterday afternoon, as the team of doctors, nurses, and specialists wearily finished up what had stretched into a two-day event, Peter Brown sat quietly in his office. He had plenty to think

about. Assuming the patient did well overnight, the hospital's first face transplant patient would wake up to a new day and a new life. So would the hospital, stepping out once again onto a high-stakes medical stage. It is Brown's job to handle that exposure.

His long night has been followed by an even longer day. A press kit, including the formal announcement pictures, is going up on the hospital's website. Two rooms have been set up: a large auditorium and the hospital's TV studio for one-on-one interviews. Security will be tightly controlled to ensure that only credentialed members of the press will be admitted, because you never know who might attempt to crash or even disrupt such an unusual announcement. As a former television news director Brown is eager to get started. It's going to be a great day for the Brigham.

But not, he knows, for the donor family. They are in mourning, ready to bury their loved one, and that knowledge pulls on him as he works.

10 a.m. Boston's North Shore.
Kay Lazar, a reporter for the *Boston Globe*, is scribbling notes at a meeting on the future of health care. She looks at her watch and wonders how much longer it will continue. Her colleague is on vacation and Kay is covering her beat, juggling multiple tasks for this proud but troubled newspaper. To save money the *Globe* recently moved its Health and Science section into Lifestyles and launched "White Coat Notes," a popular online news blog covering the city's sprawling medical community of schools, hospitals, research labs, and medical companies.

Her cell phone vibrates. It's her editor, and she leaves the room to take the call. "Kay, you've got to get down to the Brigham right away. They've just done a face transplant, and they're holding a press conference."

Kay knows the basics, but her colleague, Liz Kowalczyk, has been tracking the story, having interviewed Dr. Pomahac a couple of years ago. "Okay," she says. "I'm on my way."

Kay is not a typical reporter. Never eager to chase ambulances, she specializes in health care reform and has been busy lately with "White Coat Notes." Driving over the Tobin Bridge and into the city she tries to remember everything she can about the subject, which isn't very much. Facial transplants are not her thing. This is Liz's story, and she will be sorry to miss it. After almost an hour in traffic she stops briefly at her office to read Liz's face transplant articles before racing over to the hospital. News trucks and police cars spill onto the street as she pulls into the parking lot reserved for the press. She finds her way into the crowded conference room.

Her editor calls again, this time to tell her that right after the press conference, Peter Brown is giving the *Globe*—is giving *her*, in other words—a one-on-one interview with one of the surgeons. Great. She'll have to think of some smart questions, and fast. She wishes she'd had more time to prepare. She nods at Brown across the room, who is sandwiched between a couple of doctors. Dr. Eriksson steps up to the mike to begin the event. Focus, Kay. Focus.

chapter seventeen

Friday, April 10, 2009, morning.
Stanetsky Memorial Chapels, Brookline, Massachusetts.

I keep thinking the worst has passed, but then the next day arrives. The haunted look in Emily's eyes when she rushed into the hospital room and saw her father dead—how could anything be more awful? But today is even worse.

We are in a room. It is hard to stand up, and someone gently pushes me down onto a couch. Ghoulish pale faces with red bleary eyes are lined up next to me, like the couch where Michael Keaton sat in Beetlejuice. Jacob slumps at the other end. He is wearing his bar mitzvah suit. It fits him better now, I think, maybe.

Somebody wants to know where to sit if they're going to speak. We have a lot of people who want to speak. On the right, says the funeral director. My sister puts her head close to mine and whispers that So-and-so wants to talk with me before we start, but the funeral director shakes his head and says we must begin. He extends his hand to help me up from the couch. No, please, don't touch me.

I try to stand, and Emily grabs my waist and points me toward the door. I love having her by my side.

Susan keeps her head down, walking carefully behind Cousin Ziva, matching each of her footsteps so as not to stumble, afraid to look up and see the pain in the faces of all these people. On an easel next to the coffin is a blown-up black-and-white photograph of Joseph as a boy. He is six, dressed in a suit with a bow tie. The boy has a beaming smile and is looking up with shining eyes, as if to say, "Come and get me, you big world out there!" Susan chose this picture because it captures her husband's extravagant optimism.

A man wearing a bow tie steps up to the podium. He gestures to the picture and says that he and Joseph shared a similar taste in ties. People laugh.

Joseph's nephew, Bobby, breaks down as he describes the man who was more father than uncle to him. Others share stories about Joseph's humor, his compassion, his love of life. A friend fights for composure as he recalls their final conversation. As usual it was about family, and how grateful Joseph was to be alive when Ben opened his acceptance letter from NYU. Pam Levine, Joseph's surrogate daughter, steps up to the podium as she did at her father's funeral, ten years ago in this same room. Then Emily and Jonathan, standing together.

I can't take this. It's too much to bear. The rabbi looks at me from his chair behind the podium and nods in my direction while Emily tries to be brave. He knows the pain this is causing.

The chapel has become very quiet. But then Jonathan says, "My dad would be really pissed that he couldn't be here to listen to everyone saying how great he was." They all breathe again, and the laughter returns.

The cortege snakes its way to the cemetery, led by policemen on motorcycles in tall shiny black boots who expertly shut off intercepting traffic. As they turn a corner Susan sees a procession of cars with glaring headlights right behind.

They pull into the cemetery gates and come to a stop. Ben opens the door to get out but is told by the driver to remain in the car until everyone else arrives. Nobody feels like talking. Exhausted, Susan puts her head back and closes her eyes. She is thinking about the double plot.

She bought one in Los Angeles too, when Joseph's mother died. They brought Rachel to L.A. when her Alzheimer's got bad. A wonderful Philippine woman took care of her, and Susan and Joseph would visit every few days. But by then Rachel no longer knew her son, and she would stare at him with a blank expression. She still remembered Susan and allowed her to comb her hair and rub cream on her dry legs, until finally, one day, she no longer recognized her daughter-in-law. Susan bought the double plot thinking that when the time came she would bury Joseph next to his mother. But then they returned to Boston, and here she is with another double.

She finds it slightly ridiculous that her Polish-born mother-in-law, who grew up in a shtetl and survived Auschwitz, is buried half a world away from her birthplace, just off the 405 freeway near LAX. From a window seat on the right side of a plane she can spot Rachel's grave from the air during the final descent. In L.A. people sometimes vie for a spot near the grave of a star.

God, I hope she isn't mad at me for leaving her all alone near the 120-foot blue-tile Al Jolson Memorial Waterfall.

Twenty-eight years earlier. April 1981. Brooklyn, New York.
Joseph and Susan sail along in his little silver Datsun, the sunroof wide open on this warm spring day. New York City pops into view, and the World Trade Towers in lower Manhattan seem to lean into one another as they reach into the sky. "Joseph, I can see the Statue of Liberty!"

He's seen it enough times. The first time, although he doesn't remember it, was when he was two, coming off a ship with his parents and sister in 1951, a few years before Ellis Island was permanently closed.

They are driving to his mother's apartment in Brooklyn. Passover begins tomorrow, on Saturday night. The only thing Susan knows about the holiday is that ample amounts of bad sweet wine are consumed and no one can eat bread for a week. And just this morning, standing in line at the liquor store with a case of Kedem Concord Grape wine for the Helfgot family Seder, she learned that beer is also off limits. Susan, who grew up in a home of beer-drinking, strict German Catholics, was not aware of this. Going without beer will be a sacrifice, but it doesn't really matter. She just wants to be with Joseph in New York. Life is good, and they are in love.

They spill onto Ocean Parkway and into Brooklyn. After Naftali died it became harder for Rachel to fend off the neighborhood hoodlums. They knew she was alone, and there was little she could do when they came into the store and helped themselves to whatever they liked. In the early 1970s the family left the Lower East Side and moved into a large, post-Depression apartment on Avenue M in Brooklyn. Rachel has been there ever since.

After the store closed, Rachel's knack for making a sale landed her a job at the retail counter of Streit's Matzo Factory on Rivington Street, in lower Manhattan. She would take the train to work every morning and return home several pounds heavier at night. It was an open secret on Avenue M that you could buy Streit's products at Rachel's apartment for half of what they charged on Rivington Street.

As they pull up to the curb, Susan sees a sign on the supermarket across the street: Glatt Kosher.

"What does *glatt* kosher mean?"

"Very kosher." Joseph opens the door to the building. The halls smell of Pine-Sol, garlic, and mothballs, and the air is still stale from the long winter. Everything is painted light green. Even the linoleum floor is a dull green flecked with bits of black and silver.

An old lady at the end of the hallway is pushing bags down the garbage chute. She spots Joseph and calls out, "You call yourself a doctor? I had to buy your poor mother tuna fish today. She has vhat to eat, I ask you? Buy her some food, I *beg* you." Her voice is cracking as though she's on the verge of tears. She shuffles halfway down the hall toward them and disappears into an apartment, her slippers slapping the linoleum.

"Ignore her," Joseph tells Susan. "Bye, Mrs. Pensky," he calls out as she slams her door.

They enter Rachel's apartment. A small woman, barely five feet tall, in a black print polyester dress with white patent-leather high heels stands in the kitchen, fitting Shabbos candles into candlesticks. Her stockings are black and heavily laddered with runs. Over the dress is a yellow and orange paisley sweater, with sleeves scrunched up high past the elbows. As she lights the candles and whirls her hands over the top of the flames, Susan can see the number tattooed on her forearm. Then Rachel squeezes her eyes tightly shut and holds her hands together. Gray hair is stuffed up under a blond hairpiece on top of her head. The heels and hair add several inches to her tiny frame. She finishes the blessing and looks up at her son with a beaming smile.

"Hi, Ma," Joseph says, hugging her and handing her the bag of groceries they picked up on the way. "This is Susan."

But Rachel is busy unloading the grocery bag. "For vhat I need chocolate?" she says, holding a fancy wrapped solid chocolate Seder plate.

"Ma, stop telling Mrs. Pensky you have no money." He goes

to a cabinet and begins to rummage around, shoving aside a few large cans of tomato purée. "Susan, look." Dozens of cans of tuna fish are stacked high up in the back, along with many cans of chicken. The chicken isn't kosher, but Rachel has her own version of the rules, and chicken packed in salt water is good enough.

"*Zi nischt far du,*" she mutters under her breath.

The words sound German to Susan, and she hears the voice of her grandmother, a tall woman from Cincinnati. Susan can't speak Yiddish, but the words seem familiar. What she thinks Joseph's mother has just said is "She's not for you."

Joseph responds with a long string of words that Susan can't begin to comprehend, followed by "Did Pauline call?"

He picks up the phone and dials his sister's house. "Hi, we're here. No fish, right? Susan is allergic to fish." As he talks on the phone to his sister, Susan notices reams of heavy drapery fabric hanging from ceiling to floor on every wall of the apartment.

"For vhat you not eat fish?" asks Rachel.

"Susan is allergic to fish, Ma."

"Alloigic? I never hoid such a thing!" She takes the chocolate Seder plate over to one of the walls and pulls the fabric back, the rings on the curtain rod clinking until several feet of wall are exposed, revealing shelves filled with cans of beans, peanut butter, and tomato sauce. One section, whose shelves are lined with newspaper and doilies, holds boxes of matzoh and macaroons in tins. On a high shelf near the ceiling are shoeboxes with the Thom McAn logo. Rachel shoves the Seder plate onto the Passover shelf with the macaroons and pulls the drape closed.

"Kumn! We're going." She opens the door of the apartment.

"Ma, I don't want the candles burning while we're out."

"Kumn!" He leaves the candles alone.

Joseph stuffs his mother into the front seat of the tiny car. Susan squeezes in back. They drive to Pauline's house in Canarsie, a neighborhood filled with attached homes and fifteen-year-old cars still in active use. At a store along the way Joseph jumps out to buy milk for his sister.

Rachel has been silent since they left the apartment. Now she turns around and faces Susan. "*Schlecht blut!*"

Susan, who has no idea what the phrase means, responds with a smile. Rachel turns back and stares out the windshield.

Later that night, when they are climbing into bed in Rachel's apartment, Susan asks, "What does *schlecht blut* mean?"

"Bad blood. Where did you hear that?"

"Joseph," she says with mock innocence, "I'm starting to get the feeling that your mother doesn't like me."

"She'll get used to you."

April 10, 2009. Independent Workmen's Circle Cemetery.
"Sue, he wants us to come." The funeral director is standing outside, signaling them to get out of the car.

It feels old in this place, sacred and scented with earth. Ben goes first, scooping a bit of dirt into a shovel. He dribbles it over the top of the cherry casket. He hands me the shovel and I go next. Now everyone is taking turns. It seems to be taking a long time.

Now Mick and Dave. They loved Joseph. Their wives and kids come for barbeque in our yard, and Joseph turned them on to merguez sausage. Claude, Sharon's Tunisian husband, started the whole merguez thing back in L.A. We had Shabbat dinner with them every Friday night. We would grill while the kids swam in the pool. God, I miss her. Where is Sharon? There's Claude. Every time Joseph goes to L.A. he brings back some merguez from Claude's kosher butcher. Mick and Dave love merguez now. Joseph grills and we drink beer with shots of tequila while the kids run around. Now we won't have merguez anymore, and damn it, watching Mick and Dave is making me cry.

Mick is holding a huge shovelful of dirt. He once made Joseph laugh when he told him that if it weren't for the circumcision, he'd become a Jew just to be like Joseph. He throws the dirt on top of the coffin.

Now Dave is throwing in a heaping load of dirt. The two of them are scooping up more dirt from the pile and flinging it into Joseph's grave. They are like madmen, and I watch until they are spent, until all the dirt is gone.

Mick pats down the soft mound they have made on top of the grave with the back of his shovel, packing the dirt tight. His face is red and sweaty and streaked with tears. He throws down the shovel and walks away. I am shaking.

It is a warm, sunny day, cloudless at the noon hour. A strong spring wind blows in from the west.

chapter eighteen

Friday, April 10, 2009. Hospital press conference.

this is the kind of event a veteran newsman like Peter Brown lives for. He and his counterpart at the organ bank have been at it for two days, carefully orchestrating the announcement that was issued earlier this morning. In the auditorium, where the energy is palpable as the press conference is about to begin, Brown scans the faces and recognizes almost everyone. He was hoping for a big turnout from the regional media, and they're here. But because this is the country's second face transplant, it's been difficult to gauge the level of national interest.

Exactly four months ago Dr. Maria Siemionow performed the first face transplant at Cleveland Clinic, where she and her team replaced 80 percent of the face of a woman who was the victim of a close-range shotgun blast. Although it was never a race between the two hospitals, nobody at the Brigham will deny that it would have been nice to be first. But the second face transplant is still newsworthy, and Brown recognizes people from CNN, Fox, and ABC.

The hardest part is over; the doctors are pleased, and James Maki seems to be recovering well. The press is convivial, and members of the surgical team are in high spirits. Drs. Eriksson, Pribaz, and Pomahac mill around behind the podium with a few colleagues. A vice president of the organ bank is here too, with the somber task of reminding everyone that there would be no story without a donor, and no donor without a death.

Brown watches a member of the film crew catching a pan shot of the room. He spots Liz Kowalczyk's colleague Kay Lazar. Too bad Liz is away. He gave her the story back in the summer of 2007, when Pomahac received preliminary approval to conduct a face transplant trial. Kay's a pro, though, and Brown hopes she won't mind the documentary crew tagging along to film her interview with Pomahac after the press conference. He nods to Dr. Eriksson: it's time to start.

As Bo Pomahac waits his turn, he is apprehensive. Something could still go wrong. Maki has been out of surgery for less than twenty-four hours, and he isn't even fully awake. Pomahac won't relax for a few more days. He prays that none of Maki's new blood vessels will collapse and that there are no lurking clots.

Mid-afternoon. Helfgot residence.
The funeral is over and the house is filled with mourners. The cherry branches are beginning to bloom on the dining room table. Susan looks around for a place to put a plate of fancy macaroons, but the table marked "kosher for Passover" is completely full. The macaroons are larger than baseballs. *Who on earth would eat one of these things?*

She shoves aside a tray of lasagna with meat sauce on the nonkosher table and nestles the macaroons beside it. The basket of forks is empty again. She grabs the basket and starts toward the

Joseph Helfgot as a first grade student in 1954. A large print of this portrait was displayed next to his casket at the funeral. *Courtesy of the Helfgot family*

Joseph's mother, Rachel Wasserman Helfgot, behind the counter of the family store on the Lower East Side of New York during the 1950s. *Courtesy of the Helfgot family*

A lively Joseph Helfgot in his late forties at a dinner party in a California restaurant. It is the mid-1990s. *Courtesy of Susan Whitman Helfgot*

Susan and Joseph Helfgot on a weekend getaway at Laguna Beach in 1999. This picture ran on the front page of the *Boston Globe* when it was revealed that Joseph had been the facial donor for Jim Maki. *Courtesy of Ellen Aub Doeren*

Joseph Helfgot with his wife and four children at Chef Chang's, a favorite Chinese restaurant near their home in Brookline, Massachusetts, during Thanksgiving weekend in 2008. He died five months later. *Left to right:* Susan, Benjamin, Joseph, Emily, Jonathan, and Jacob Helfgot. *Courtesy of Joan Ganon*

Joseph Helfgot with his son Benjamin in September 2008. A tracheostomy tube aids his breathing and an implanted ventricular assist device helps his heart to beat. The VAD wires embedded deep in his abdomen connect to external batteries hidden in pockets sewn into the lining of a vest. *Courtesy of John D. Brink*

Jim Maki in his backyard at his childhood home in Seattle, Washington.
Courtesy of Jim Maki

Jim Maki in Massachusetts shortly after returning from Vietnam in 1970.
Courtesy of Cynthia Maki

Jim's father, John (Jack) Maki, at age twenty-one, shortly before graduating from the University of Washington.
Courtesy of the Maki family

Newlyweds Cynthia and Jim Maki at Hilton Head, South Carolina, in the late 1970s. *Courtesy of Cynthia Maki*

Jack and Mary Maki playing with baby Jessica. *Courtesy of Cynthia Maki*

Jack Maki with granddaughter Jessica in the mid-1990s, taken at his home in western Massachusetts. *Courtesy of Cynthia Maki*

Jim Maki with Jessica at her high school graduation in 2004. *Courtesy of Cynthia Maki*

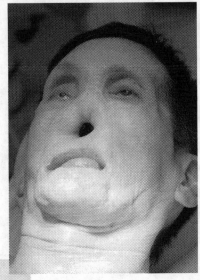

Jim Maki awaiting surgery on the evening of April 8, 2009.
Courtesy of Dr. Bohdan Pomahac

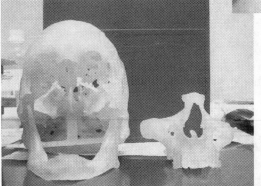

Exact acrylic models of Jim Maki's disfigured head and the portion of Joseph Helfgot's face necessary to restore it sit side by side on a table in Bohdan Pomahac's office.
Courtesy of Dr. Bohdan Pomahac

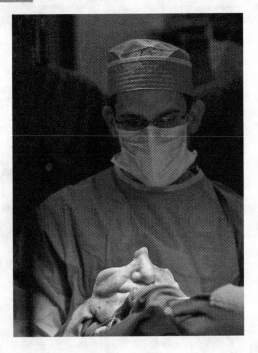

The vascularized composite allograft—skin, blood vessels, nerves, bones, and muscles—moments after being surgically removed from the donor, Joseph Helfgot. Tagged arteries and veins are visible on the lower left.
© *April 9, 2009 by J. Kiely, Jr., Lightchaser Photography*

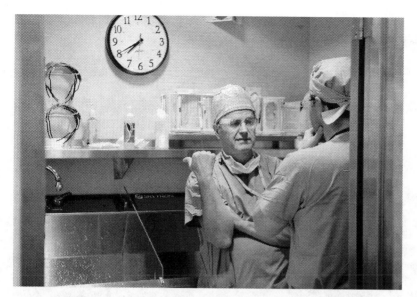

Julian Pribaz discusses the second half of the surgery with fellow surgeon Bohdan Pomahac, his former student at Harvard Medical School. Pribaz motions toward the Maki operating room, where a team is enlarging one of Maki's arteries to better match up with its donor counterpart. The delay provides a brief rest during the 17-hour surgery. © April 9, 2009 by J. Kiely, Jr., Lightchaser Photography

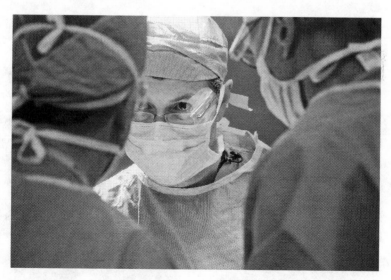

The intensity of the moment is captured in Dr. Pomahac's eyes as he prepares to sever the final blood vessel that will transform a portion of Joseph Helfgot's face into an organ transplant for Jim Maki. Elof Eriksson, Pomahac's mentor and chair of Brigham and Women's Plastic Surgery Department, assists on the right. © April 9, 2009 by J. Kiely, Jr., Lightchaser Photography

Jim Maki and Susan Whitman Helfgot appear together at a hospital press conference six weeks after transplant surgery, on May 21, 2009. *Courtesy of the Department of Public Affairs at Brigham and Women's Hospital*

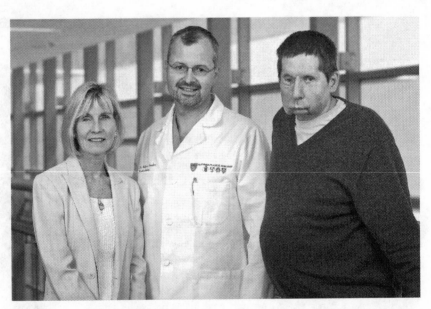

Susan Whitman Helfgot, Dr. Bohdan Pomahac, and Jim Maki thirteen months after Maki's transplant surgery. © *May 13, 2010 by J. Kiely, Jr., Lightchaser Photography*

kitchen. She notices Esther from the organ bank, who is talking to a doctor from the Brigham. She is glad that Esther has come back to the house.

Dave, who shoveled dirt on Joseph's grave, has been handing out pins that look like campaign buttons. A few years ago he and his wife attended a karaoke party with Joseph and Susan to raise money for their children's school. Someone snapped a photo of Joseph singing with the mike in his hand, and Dave has hand-pressed hundreds of buttons with that picture on them, using an old-fashioned machine in his basement. Ben squeezes past his mother in a tan ski cap that he hasn't removed since his father died. There are Joseph buttons all over it.

The kitchen is filled with visitors. Several of them are holding drinks and watching CNN on a TV set on the wall.

"Hey, look," says one of the mourners, who has already consumed several shots of tequila. "They just transplanted some guy's face. Can you believe it? Like that chick whose face got blown off by her husband." Everyone looks up at the TV. Someone in a white coat is speaking. It must be one of the surgeons.

Someone else steps up to the podium, a man from the organ bank. Susan starts to tell her sister to find Esther, but she stops herself just in time. Everyone would know. She listens along with everyone else. "Advances in transplantation only happen when there are individuals and families who can see past their own tragic circumstances and agree to donation . . . It was New England Organ Bank's honor to work with such a remarkable donor family."

They cut away to a video of a surgeon holding a translucent plastic jigsaw model of the facial allograft in one hand. On the table is a head made of the same material, and Susan watches the surgeon insert a stand-in for her husband's face into the plastic skull. They cut back to the doctor who is answering questions from the podium. Susan thinks she will faint. "He's still waking up," the doctor is saying. "He hasn't seen himself yet."

The TV anchor says, "The donor family wishes to remain anonymous, but they have released the following statement: 'To go from being a recipient family to a donor family so suddenly has given us the opportunity to fully understand the power of organ transplants to give and transform lives.'"

It sounds false as I hear it, like a bad actress reciting a lousy script for the first time. And why is she reading only one sentence? Jon and I spent almost an hour composing the statement. Next time I'll write just one line. Wait—next time? What am I thinking?

"Great story, isn't it?" responds the coanchor, smiling widely. Susan's friend from California, who knows what really happened, looks over to her. Susan winks, hoping no one notices how red her face has become.

The hospital.
"Kay?" Brown taps her shoulder. "Dr. Pomahac will be happy to speak with you now. We have some people in town who are filming a documentary. You don't mind if they come along?"

Kay goes down a hall, distracted by the cameras, trying to compose her first question. When she is introduced to Dr. Pomahac, she sees a tired young man, and the questions swirling in her head fade away. "How does it feel?" she asks.

"Surreal."

Helfgot residence.
"Sue, is there someplace we can speak privately?"

Susan leads the general manager of Marketcast into the library and closes the door. Henry Shapiro has had a hand in running Joseph's company almost since the day Joseph and his partners sold it to Reed Elsevier several years ago.

"It's not going to be the same without him," he says.

"I know."

After Joseph and Susan moved the company to L.A. in 1996, Marketcast grew quickly. By 2000 two suitors were vying for it. As a major supplier of business-to-business information, Reed Elsevier wanted to be the top player in providing market research to the lucrative entertainment industry and had already bought *Variety*, an entertainment daily.

The Marketcast acquisition got off to a poor start when the Dutch conglomerate tried to nudge Helfgot aside. Marketcast's president was fun to hang around with, but he was also a loose cannon, saying whatever came into his mind at corporate meetings and arguing with senior management about almost everything. He was also expensive. It wasn't just his salary, but his lavish tastes. Even on short flights he refused to fly coach. He was known to run up five-hundred-dollar business lunches by dragging along favored employees who would normally be munching on sandwiches in the corporate lunchroom.

The executives at Reed were under the impression that movie studios were buying Marketcast's services for its sophisticated statistical analyses of moviegoers' preferences. But they quickly learned that what the studios really wanted was someone who would let them know straight out what the hell was going on with their movie, good or bad. They trusted Helfgot to tell them what all the numbers on those charts really meant, and to be frank with them when they didn't mean anything. Honesty can be scarce in Hollywood, and Joseph Helfgot was brutally honest. It was his greatest asset, although it regularly got him into trouble.

Henry Shapiro was tapped to bring some order to Marketcast, and those who planned Joseph's early demise were soon gone. Shapiro's business acumen blended well with Helfgot's marketing insights, and the company enjoyed explosive growth, its revenues increasing dramatically in just a few years.

"We've been talking it over, Sue, trying to come up with something Joseph might like to be remembered by. We're thinking of endowing a scholarship at UCLA in his name, in sociology. Do you think he would like that?"

Susan is stunned. "I think he would."

Twenty-five years earlier. Cambridge, Massachusetts.
"How'd you do?"

"Not too good." Joseph has just come home after his weekly poker game with the guys, several of whom are independent theater owners. "But I may be able to get into General Cinema. J.D.'s wife told me about a market research study they've just finished. It's completely flawed. I can't believe they paid so much money for it."

For the past few years Helfgot has earned some extra income outside of his teaching job at Boston University by taking on small market research projects around Boston. Fidelity, Lotus Development, Wang, and Polaroid have all hired him to shed light on big decisions, such as the color of a new camera for kids.

"I think Dick Smith likes me," Joseph says after his meeting at General Cinema. "I told him that video rentals were going to be the best thing that ever happened to the movie business." At the time, the exhibition industry was terrified about the impact of VCRs on their business. If people could rent a film and watch it whenever they liked, without worrying about babysitters, ticket prices, or anything else, why would they ever go out to the movies?

Helfgot had a different view. He believed VCRs would actually save the industry. Baby boomers with young children had no time to go to the movies. But if they could rent movies and watch them at home, after the kids were in bed, they would remain interested in what Hollywood was producing and would return to the theaters in a few years. So would their kids, who would be growing up in homes where movies were as familiar as television.

"I told him they needed to do more segmentation on which films they should put in a given market, and that we could poll local audiences. Maybe they could start some sort of frequent moviegoers' club. Then they could segment the avids from the occasionals and the rares." Joseph had already thought about the different kinds of moviegoers. "I'm not sure they like that idea, but who knows?"

Four years later. May 15, 1988. A private club in Boston.
An hour before Susan's wedding, her sister is putting on the bride's makeup. Susan is so tired she can barely stand up. She spent the night stuffing five hundred brochures into envelopes, which she ran over to the post office early this morning. Joseph intended to help, but Rachel was spending the night at their house and she ordered her son to get some sleep. The brochures highlight a study just completed by Helfgot and his two colleagues from the State University of New York at Stony Brook, "Aging Baby Boomers and Declining Leisure Time: Strategic Implications for the Movie Industry." They hope to attract a few nibbles from studio executives and get a foothold in Hollywood.

On their honeymoon in Antigua Susan calls the office. "Anything come in yet?"

"You have two return cards."

"Joseph, we got two cards!" But two trips to L.A. and a thousand dollars in credit card bills later, both leads have fizzled out.

Joseph is undeterred. He continues to consult to General Cinema and then to its rival AMC. At Helfgot's urging AMC launches a program called MovieWatcher, a club for avid filmgoers, which is still going strong.

One day, out of the blue, Orion Pictures calls. They are worried that one of their epic movies is going to flop. Someone has told them about a quirky guy in Boston, and the head of marketing remembers Helfgot from a trade show, where he was brilliant but unfocused. What the hell, they've got little to lose. Kevin Costner in a very long western is going to be a tough sell.

"A western? Three hours long? With *subtitles*? Sure, I'll be there this afternoon."

"Orion wants me to come to New York," he tells Susan.

"Honey, that's great."

"Not great. A three-hour western with subtitles. Are they kidding? No one will see that. I'll give them advice, the movie will bomb, and that will be the end of it. Would *you* see it?"

"I like Kevin Costner."

"Let's hope he likes me. He's directing this thing, and he's supposed to be at the meeting."

"Here's what you need to do," Helfgot tells the Orion executives. "Women love Costner. Forget about the main story line. This is a movie about a man in search of himself, who is struggling to be good. Women love that shit. But he's really brave too, a guy's guy. That's crucial. It's the only way you get the couple's butts *together* in the seats. If she wants to see it, because of Costner, he won't go and they'll see something else. And if you try to sell it as a guy's movie, she won't want to see it. She'll stay home with her girlfriend and he'll go bowling or watch football with a buddy, because guys won't go to a western together. This isn't *The Terminator*. You need to get couples to go *together* or you're dead."

The executives sit around the table as Joseph paces back and forth with a dry-erase marker in his hand. *Dances with Wolves* is just about in the can. It's time to decide how to position it.

One of them sighs deeply. "Okay, let's go with it." At Helfgot's suggestion they market the movie differently in different parts of the country. The East and West Coasts will see trailers with a Native American slant, and the middle and southern areas will get the American soldier slant.

"Suze, what do you think?" Joseph shows her a mockup of the poster. The caption under Costner's face reads, "Lt. John Dunbar is about to discover the frontier . . . within himself."

"I like it."

"They really got it. They got it *exactly*."

"Joseph, stop crying."

Dances with Wolves won seven Oscars in 1991, including Best Picture. At the time, Oscars were not given to movies that nobody saw.

The bedroom phone rings. "Joseph, what do think about a movie starring a guy who eats people?"

"I think you're out of your mind. It sucks." He hangs up the phone. "Suze, I'm going to Orion. They've got something about a cannibal and an FBI lady."

Susan rolls her eyes. "Good luck with that."

Joseph's focus groups tell him that after *Taxi Driver* and *The Accused*, audiences see Jodie Foster as a victim. They don't believe her character can stand up to Hannibal Lecter, a serial killer who targets women. So the studio adds two scenes to the trailer. In the first one, Lecter, locked in his prison cell, terrifies her, and she backs away. In the second, when he tries to intimidate her again, she holds her ground and Lecter backs off. The movie opens on Valentine's Day, although there's nothing romantic about it. It becomes a huge success. A year later *The Silence of the Lambs* wins five Oscars, which doesn't normally happen with horror thrillers.

Twelve years later. Spring 2000. Santa Monica, California.

"Joseph, what's happening?" Susan is in the car, stuck on I-10, heading home toward the Palisades. Joseph has called her three times today. He's been in meetings with the lawyers to finalize the sale of the company. She has kept her cell phone in a sweaty palm since leaving for work this morning, afraid to set it down for a minute. If this deal goes through, they're set for life. It's hard to believe they had all of fifty-seven dollars in the bank when they got married. They were tapped out, juggling their two businesses and trying to make their payrolls every week. Now, barring a last-minute glitch, they can finally relax a little.

Four years ago they sold Susan's financial services company and their house in Boston and moved to L.A., sinking every dime they had into Marketcast. Joseph opened an office in West L.A. on the Santa Monica border. Two employees came with him from Boston, and he hired several more. But employees are expensive, and so is everything else. Rachel, who died recently, had been in a nursing home. There were college loans for Joseph's older kids, and two younger children to care for. Susan can't believe they had the chutzpah to think they could pull it off.

"Is everything still okay?" she asks. She worries that if the deal falls through it will kill him. Last year Joseph spent several weeks in the hospital after passing out at lunch. They rushed him to UCLA Medical Center, where he was cared for by Gregg Fonarow, a young cardiologist who was trained back east by Dr. Lynne Stevenson at the Brigham. Dr. Fonarow told Joseph he would soon need a new heart, and Susan attended a "What to expect when you're expecting a heart" meeting for spouses of patients. Dr. Fonarow started Joseph on a new regimen of drugs, and the patient seems better now.

"Suze, I'm holding the check."

"Oh, Sweetie."

"I'm at the bank. You know what? The number is too big to fit on the deposit slip. I wish my mother could see it."

"Honey, I'm so proud."

"Meet me for a drink?"

"Twenty minutes."

When Susan arrives at their local hangout Joseph starts crying. "I keep thinking my mother should be here. Your dad too." Susan's father is dying of a brain tumor. For eight months she has been flying back and forth to South Carolina to help her mother. Now they are both crying.

They walk out holding hands into the late afternoon sun and stand by their cars for a moment. Before Joseph opens the car door, he turns around and shouts out, "Unf—ing believable!"

Susan giggles and gets into her car. She follows him home, talking with him on her cell phone the whole way.

April 10, 2009. Helfgot residence.
Someone bursts through the library door, a studio guy from Hollywood. He's holding a pair of jeans in one hand and sneakers in the other. "Susan, I can't stay in this suit another freaking minute. Can I change in here?" He starts unbuckling his belt.

"Jeez, keep your pants on. Let me out of the room first."

"I just keep thinking," he says, pulling off his dress shoes, "who am I going to call now?"

chapter nineteen

The media people pack up their equipment, eager to get back and edit their stories. One of Maki's doctors makes his way over to Dr. Kim.

"Congratulations," she says. "How is Mr. Maki doing?"

"He's awake. Go take a look."

She makes her way into the ICU and looks around for her patient. To ensure privacy, Maki's name is not on the door. The desk nurse nods in the general direction of his room.

From a distance Dr. Kim makes out a facial profile. She is wildly excited and takes a deep breath. Normally there is nothing there when she looks at James Maki from the side—no contours where a mouth or nose would be, just one flat, featureless surface. But now, even from far away, she sees something very different. His face is grotesquely swollen, but there it is: an honest-to-God *face*.

She walks quietly into the room and stares down at the new James Maki. "Mr. Maki?" she calls out softly. No response. Half-sedated, half asleep, he doesn't stir. "You look great," she whispers before she leaves.

Nineteen years earlier.
University of Massachusetts at Amherst.
The seventy-year-old professor quickly makes his way across campus to his office. Spry for his years and looking much younger, Jack Maki is ruminating on his future. He will retire in a few months and will sorely miss academia. It has been his life for half a century, and he finds it hard to imagine any other existence. His department has generously offered to let him keep his office for as long as he wishes, but he has watched other retirees hang around longer than they should have, injecting their opinions and gradually turning into unpopular has-beens.

No, he'll give up the office. So much of his life is on display in this cluttered room—all his awards, pictures, degrees, and, of course, his books.

He has achieved quite a lot for a man who was abandoned as an infant. He sees a picture of his family, and his thoughts immediately turn to Jim, who came to the house yesterday, once again begging for money.

He had such hopes for the boy. He remembers the day Mary received a call that a half-white, half-Japanese baby was available. They had already adopted John and were looking for a girl. But Mary convinced him to go and see the baby. They named him James. He had hoped things would improve when they moved to Amherst. The small college town in the foothills of the Berkshires seemed to offer the hope of a quieter existence. Maybe Jim would settle down and focus on his studies.

• • •

June 1966. Seattle.

"Boys, we have some news."

Jim and John stopped eating and looked up from their plates. Their parents never had any news.

"I have been offered a job to start an Asian studies program at the University of Massachusetts next fall." Jack Maki raised his chin a bit higher and smiled at his family.

It didn't quite sink in at first. "That's good, Dad," John said.

"We'll be moving in July."

"*What?*" Jim gasped.

"In July," his mother said. She looked happy. Mary Maki was always happy when her husband was happy.

"Mom, we can't move. I have to graduate."

"You'll graduate in Amherst. They have a very fine high school," his father explained.

"Well, I'm not going. You can't make me." Jim jumped up from the table and threw down his fork. He ran up the stairs and slammed his bedroom door.

July 1966. Western Massachusetts.

Heavy rain pelts the roof of the white Galaxy 500. Jack Maki leans over the steering wheel, trying to see through the windshield in search of an exit sign.

"This is the one." He eases off I-91 onto a country road.

In the backseat John opens his eyes. Six long days of driving. Will they ever get there? Jim sits next to his brother, watching their father. He drives like an old man, his knuckles white as he grips the wheel.

Dr. Maki has always seemed old to Jim, at least compared to their friends' fathers. Jack feels it too. He even subscribed to *Sports Illustrated* in the hope that it would bring him closer to Jim.

Jim stares out at the New England countryside, which is filled with rolling hills covered in pine trees, green humps on a camel's back. It's so different from Seattle. It's hot in the car, and his father is using tissues from Mary's purse to wipe the condensation off the windshield. The wiper blades barely keep up with the heavy downpour.

"We're almost there," his mother says. "Isn't this exciting!"

Not really. Jim mouths the words silently to the back of his father's head. John stifles a giggle. The boys aren't that close, especially since John started college last fall. The constant tension between Jim and his dad hasn't helped, but this trip has brought the brothers closer. Their world has been turned upside down, more for Jim than for John, who will head back to college in Seattle next month. John feels sorry for his brother, who will have to deal with a new high school during his senior year. That's going to be tough.

The rain abates, changing to a fine mist. Jim cracks his window. Sweet air flows into the car, different from the air in Seattle. Earthier. *This is where I am going to spend my senior year in high school? In the middle of goddamn nowhere?*

John shuts off the engine and they stare at the rental house from the car. It is small and looks abandoned. Newspapers scatter the walk, soaked from all the rain.

The humidity is oppressive as the storm moves off toward the east. They sit in the car for a moment without speaking. Jim has never experienced silence this complete. It's as if the four of them are the only people on the whole planet. A hard lump forms deep in his throat.

"Well, what do you think, boys?"

Jim pulls on his duffel bag and gets out of the car, slamming the door. He heads up the walk to the house.

"Where do you think you're going?" his father calls after him. "Come back here and help us with the bags."

"Maybe he needs a little time," Mary says. "It's a big adjustment."

• • •

1990. University of Massachusetts at Amherst.

That was more than twenty years ago, Jack remembers now, sitting in his office. Where did the time go? They've had a full life, he and Mary. Few real struggles or tragedies. Except for Jim.

Jack hoped things would improve when Jim married Cindy in 1978. Her father was a manager at the General Motors plant in Framingham and helped him get work there. But Jim was fired for using drugs on the job. It was that damned Vietnam War that got Jim on drugs, but Jack didn't realize it until a year or two after he returned.

Jim had done miserably at Amherst High and graduated near the bottom of his class. He had no interest in college, but when Jack insisted that he couldn't live at home unless he stayed in school, Jim enrolled at Greenfield Community College.

On his first day there he signed up for every intramural team they had. He barely had time to make it to class. And then he found a new sport. Jack shakes his head as he thinks about it. Another pointless obsession. Golf.

Spring 1968. Cherry Hill Golf Course, Amherst.

Jim surveys the hole and pulls his driver from his bag. It's a long par 5, with water hazards at the best lies. Tee off too hard and the ball may land in water; too soft, and forget a birdie. The last hazard abuts the green on the approach. Tricky.

He should be in class, but it's a mild spring day, which in New England isn't something to take for granted. Nobody is going to begrudge a guy a game of golf on a day like this. Birdie for sure. He steals one last look down the fairway and checks his wrists, elbows, and feet. In one seamless motion he brings down his left shoulder to meet the tee, sending his arms high into the air. Then, reversing his weight, he brings the club down hard, connecting with the sweet spot in a loud clean crack. The ball lifts very slowly, maximizing the distance.

Two caddies are playing ahead, but now they hang back, watching him. He's going for a birdie. Jim decides on his 3-wood, taking the left approach. Crack, and the ball flies over the next two hazards, coming to rest slightly in the rough. The grass isn't too deep, but it's spongy. He takes his 5 and looks for the flag, but it is obscured behind a bush. One of the caddies runs back and spots it for him. Jim waves thanks and brings the ball cleanly out of the rough. There is no wind today, and it clears the last water hazard, bouncing sharply, rolling downhill past the cup, coming to a stop several feet past the hole.

"Five bucks he sinks it," the caddy says to his buddy. Young Jim Maki has already established himself as a very talented golfer.

Jim assesses the green. It's a nice soft rise up to the cup, a 14-footer, give or take. Not impossible. He hears a car engine start up, but otherwise it is completely still. He loves this part of the game. It's not like basketball, where everybody is screaming at him from the stands, "Shoooooooot!" After a basketball game he can hardly hear for an hour or two. With golf, you take your time. When a player putts, everybody stays silent. It's just you and the ball.

He checks the hole, repositions himself, and gently sends the club through the ball. It runs up to the cup, slowing gradually until it stops half an inch short. He picks up his gimme.

"Tough luck, Jim," the caddy says. "I thought you had the birdie."

"Next time." He smiles to himself. It's only April.

chapter twenty

ell me you didn't." Dr. Maki grabs Jim's arm.

"Yep, I signed up today."

"Son, do you have any idea what you have just done?" Jim's father is horrified. Mary is pale. She sinks down onto the couch.

"I was going to get drafted anyway. I might as well go on my own terms."

"You might not have had to go at all if you had just stayed in school. You're quitting."

"What am I quitting? Dad, I'm going into the army. There's a *war*."

"Jim, U. Mass has a place waiting for you any time you are ready."

"I'm not going back to school. Anyway, it's too late. I've already signed up."

They took him to the airport a few weeks later. "Call me the minute you get settled in," his mother implored, hugging him. "I'm so proud of you, Jim."

As they drove back to Amherst, Jack comforted his crying wife. "You know, Mary, military discipline may be good for him. Maybe it's what he needs."

May 1969. Fort Lee, Virginia.
A few hundred young recruits stand at attention in the hot sun. "Okay, men. You are now considered one hundred percent AIT— advance infantry trained. I won't lie to you. Most of you will go to Germany, but some of you are going to Vietnam." A sergeant goes through a roll call handing out orders, but Jim's name is not called. There has been a mix-up, but the sergeant assures him he will be going to Germany. He calls his parents to tell them the good news.

That afternoon he heads over to the office to pick up his orders. The desk sergeant checks some papers and then hands him a form. "I got bad news for you, Private. You're going to Vietnam."

Several months later Jim's brother is drafted. Mary is inconsolable. A few months later, when he finishes basic training in South Carolina, John receives orders to leave for Germany. They won't send a family's only two sons to Vietnam at the same time. John will be forever grateful to his younger brother for perhaps saving his life.

A few months later. South Vietnam.
Pvt. James Maki looks out over the country that will be his home for the next year. They should be landing any minute. In the distance he can see the runway strip at Biên Hòa Air Base, where planes full of fresh troops dump their live cargo to sit and wait for orders. Suddenly the plane rises higher and begins to circle. The base is under attack and they can't land. The men make

small talk about the oppressive heat while the plane cuts lazy loops high above the base. Their eyes peer through small windows as the rockets explode below them.

No one will admit it, but they are scared. Jim looks around at the faces of the other men. He signed up; many of the others were drafted. But how they got here makes no difference at a moment like this. The same fear touches them all.

They stay at Long Binh near the air base for a few days. It's a nasty place, dirty, and poor beyond anything they have ever seen. It is one big dirt pile filled with desperate Vietnamese living in squalor, scratching it out while the war drags on—badly, ever since the Tet Offensive.

Several days later he receives his orders and is sent to Phu Bai, a small village at the start of the conflict that has rapidly expanded into a sprawling town built of sandbags. Much of the labor is supplied by children, some as young as eight, who earn a dollar and a half a day standing in the broiling sun, filling up bags.

Heavy construction supplies move through Phu Bai. So do other things, like toothpaste, soap, and cigarettes. A small pack filled with a soldier's toiletries can bring seventy-five dollars on the black market. Pilfered cartons of cigarettes go for close to thirty. A soldier can make a killing in Phu Bai, and some do. What appears on the surface to be an efficiently run facility is more like a western frontier town.

Tempers run high in Phu Bai, and the command hierarchy always seems on the verge of imploding. There is no reason to suck up to those above you, because they too want to get out of here.

Shortly after his arrival Jim stands in a makeshift hospital on an errand for his commanding officer. A North Vietnamese soldier in army fatigues is half propped up on a gurney, moaning. He seems badly hurt. Two Americans are standing over him, shouting in his face. They want to know how many others were traveling with him when he was captured. The man is writhing in pain. One of the soldiers begins to slap him hard across the face without letting up. Jim knows these tactics could save the lives of soldiers who are out combing the nearby roads, but he is sickened by the sight.

When he first arrived he was told to head for the bunkers the minute they came under missile fire. The first time the sirens went off, he and a buddy did as they were told. They were the only ones there. Twenty minutes later, when the all-clear signal was heard, they were still the only ones in the bunker. As they came out into the daylight they saw men sitting on the rooftops of the barracks.

"You idiots, what were you doing down there?" someone shouted. "No one uses the bunkers." A few guys on a nearby roof started laughing at them.

The Viet Cong are efficient. They train their missiles on the hospital and the fuel dump. They don't waste rockets on the barracks, which are too spread out. When the men hear the siren they hop up on the roofs and watch the spectacle, smoking cigarettes until it's all over and they are forced to return to whatever dull job they've been doing. Like a fire drill at school, the rocket attacks break up the monotony. They're a lot easier to take when you know they're not aimed at you.

Except when they are. Jim awakes with a start in the middle of the night. He has somehow fallen onto the floor, across the room from his bunk. But his bunk is gone, and so is the wall. A stray rocket must have landed nearby, and the shock waves sent him flying. Five feet closer and he would have been killed.

Jim doesn't do much of anything until he is finally assigned to be the driver for a senior officer. The officer travels around, supposedly checking up on security, but mostly going wherever he wants. Jim drives him through the small shanty villages that form and then quickly evaporate, depending on where the fighting is.

The villagers make temporary shelters in holes they dig into the ground, covering their "hooches" with leftover cardboard that serves as both walls and roofing.

Driving around is not that bad. It kills time, and the steady breeze in the open jeep is a welcome relief from the constant damp heat. The roads smell a lot better than the villages, which are rank with the odor of rancid fish sizzling on small tin grills.

Spring 1970. South Vietnam.

Private Maki's driving job didn't last long. The captain was reassigned, and they still weren't sending him out to work on anything. Jim hung around camp all day, playing cards and dice. The men played for money, sometimes a lot of money. Why not? Life was short and brutal here, and nobody cared.

One day Jim laid into a buddy who owed him close to four hundred dollars from playing dice. A little while ago, when Jim was on the losing end, this guy had been ruthless about collecting. Now Jim was returning the favor.

"I want it. Now."

"You'll have to wait. I don't have any money."

"I want it right now."

The soldier slipped him a packet of white powder. "This is all I got."

"What the hell is this? I want my money."

"It's all I got. You want it or not?"

Maki wasn't sure what was in the baggie. Heroin was cheap here, not much more than a pack of American cigarettes on the black market. And it was easy to find. Mama-sans, the madams in the villages, always had some lying around. No one was watching you score. No one cared.

Jim had never done drugs, but that night he learned how. He and his dice buddy got extremely high. Jim wanted more. It took a few weeks for the man to repay his debt, but Jim's fondness for heroin didn't end there.

Late one night he returns to barracks. A large sergeant who has been hassling him since his arrival shouts at him. Jim shouts back. It's hot and filthy and they both hate this place. It takes several men to separate them.

"Maki, he's a *sergeant*. What on earth were you thinking? You can't punch a staff sergeant. I don't care what he did to you. He's allowed to beat the crap out of you. You're confined to barracks for a few days while we sort this thing out."

In any other war Private Maki would be court-martialed for having struck an officer and likely sent home to face dishonorable discharge. But this is Vietnam.

"We've decided to transfer your sorry butt out of here. Do you have any idea where you want to go?"

"Maybe on leave?" he says hopefully.

"Don't get smart. You've got ten seconds."

"First Cav?" First Cavalry has a rep for the most kills. It's where you want to be in a war, with men you can count on.

"First Cavalry, Eighth Engineers. You ship out tomorrow morning. Get your gear in order."

"Thank you, sir."

The First Cavalry has been operating just northeast of Biên Hòa, shoring up defenses while it ferrets out enemy infiltration routes. It's an extremely dangerous place. Maki is supposed to be training as a combat engineer, someone who can build things even while short-range missiles and bullets fly overhead. But his new unit leader won't let him go on any missions. He knows why Maki was transferred and isn't sure he can be trusted. Out in the jungle with the enemy, a man has to know his back is covered. He'll watch the new kid for a while and see what he's made of.

• • •

Sunday, April 12, 2009, morning. North shore of Boston.
In Cynthia Maki's kitchen her striped tiger cat, eager to be fed, threads its sinewy body between her ankles. Cynthia prepares a bowl of food and sets it on the floor. It's time to get dressed for the trip into Boston to visit Jim.

Jessica is already dressed. She has never been the kind of person who will keep you waiting. Her college graduation is three weeks away. The money for school came from her aunt Michi, Mary Maki's sister. It saddens Cynthia that Michi is in far-off South Carolina, too old and frail to come north for Jessie's graduation. And Mary, who would have been so proud, has been dead for almost twenty years. Her ashes were scattered from a footbridge over the Ohana Pecosh Falls in Mount Rainier National Park by Jim's father and brother. Dr. Maki took a picture of the white powder swirling in the breeze; Cynthia used to look at it when she visited him. Almost five years ago John returned to the same footbridge and scattered Jack's ashes into the flowing waters.

Jessica's middle name is Mari, for Jim's mother. She is getting a bachelor's degree in communication. Jim won't be there either. Just Jessica and Cynthia. After graduation Jessica is going to Europe with her friends. It doesn't seem to faze these kids, getting on a plane and going wherever the winds take them. Jessica has always been confident, just like her father. The audience gave her a standing ovation after she sang "Sit Down, You're Rockin' the Boat" in her elementary school's production of *Guys and Dolls*. Once she got a child's role in a play at the North Shore Music Circus, where Stephanie Mills played the lead. It was a small part, but performing is in her blood.

Cynthia used to sing and dance. Her dance teacher had been a Rockette, and Cynthia dreamed of becoming one too. But she was an inch too short to be considered.

In the summer of 1968 Cynthia and her friends drove to the Newport Folk Festival, where a woman named Janis Joplin sang in a way Cynthia had never heard before. Back at U. Mass she started singing everywhere she went, trying to sound like Joplin. One day a guy asked if she wanted to sing with a group called Clark, Walter, and the Alligators. Later another band wanted her, but Cynthia was in the middle of a semester. She was proud of herself for sticking it out in school. Her mother didn't believe she had really graduated until Cynthia showed her the diploma. Her parents hadn't gone to her graduation.

She sits at the kitchen table for a minute before getting dressed. She likes her kitchen, which is brand-new, with white cabinets and black counters. She has worked hard to get this place. She hasn't had an easy life, and it hasn't gone the way she expected. She and Jimmy have been married for thirty-one years. Unbelievable. They were married in 1978, during March Madness, when Jim was watching one of the games on TV. At halftime they ran out and got married. They had been planning a small wedding, and this wasn't exactly eloping, but still—at half-time during a basketball game, at the house of the local magistrate? It seems crazy when she thinks back on it.

Jimmy loved watching those games. He was a really good basketball player too. And softball, and golf. He could run a pool table several times in a row. In every sport he tried he seemed to be a natural. She can still see him playing shortstop for a slow-pitch softball team in Amherst with a cigarette in his mouth. With his eyes half-closed, he'd scoop up a fast grounder and fire it to first like it was nothing. The memory makes her smile.

On Wednesday Jim called her on his way to the hospital, but she was at work and heard the message only later. On Friday she watched the news about the face transplant. It was strange to hear people talking about her husband on television, but of course they didn't mention his name. She didn't realize how big a deal it was until she saw Dr. Pomahac on CNN.

She still can't believe Jim survived the fall. She sat in a room full of doctors, discussing whether or not he would even survive the week. The poor guy was hurt beyond recognition, his face gone, his arm and hand badly burned. At best he would need as many as a dozen surgeries just to fix his face, Dr. Pomahac told her. And it wouldn't really be fixed, even then. He might need a feeding tube for the rest of his life, and a tracheostomy tube as well. They weren't sure he could breathe on his own if they turned off the respirator. But they had to try, and he started breathing through the tube on his own. First the aneurysm, then the subway fall. Jim was like a cat with nine lives.

"Jessie, you ready to go?" Cynthia wonders what her husband will look like with his new face. He used to be so handsome.

A single shot rings out. There was no warning of enemy activity in the area. It came from right outside the perimeter of the camp. They sit straight up on their cots in silence, waiting for another round. Jim's heart is pounding.

But it's quiet. They hastily gather together and make a plan, fanning out in an arc, working their way beyond the fringe of the base, toward the general direction of the sound. Jim takes slow, careful steps, his rifle in his hands, until he comes to the top of a crest running along a deep ditch. He approaches it with caution. The sound came from very nearby. He inches forward, closer and closer, craning his neck barely enough to see down into the hole. He quickly draws back. "Here!" he shouts. "I see something."

They gather around the top of the ditch. At the bottom lies a fellow soldier with a single shot to his head, his eyes wide open, his legs and arms splayed at unnatural angles around his torso. He has killed himself with his M-16. The rifle lies next to him in the ditch. As Jim bends over the site, someone taps him on his shoulder from behind. He jumps.

April 12, 2009. The hospital.

"Mr. Maki?" Nurse Lorrie is shaking him gently out of his recurring dream. "Wake up. You have visitors."

He looks up. Cindy and Jessica are standing by the bed.

"Jessie? Hi."

"Dad?" Jessie begins to cry.

"Jess, please don't cry. How do I look?"

"Really good, Dad. I never thought you'd look like this again."

chapter twenty-one

a skinny young man with an enormous Jewish Afro stands outside the Manhattan Draft Board office, smoking his third cigarette. He paces nervously back and forth, rehearsing and re-rehearsing in his mind the list of things his Quaker draft counselor has told him. He looks down at his shoes and checks the heels. They are caked with mud. He throws the cigarette onto the sidewalk and grinds it out. Then he bends over and swipes at the ashes with his hand before rubbing his dirty fingers over his shirt and neck.

As he enters the building and passes the MPs, he utters a silent prayer. From out of nowhere the image of Naftali Helfgot appears to him. He rarely thinks about his father, the proud and crippled Auschwitz survivor who was beaten down by the terrible things he had witnessed and endured. Naftali never succeeded in America, and Rachel constantly berated him for his lack of business sense. Marginalized by his wife, he hung around the store, schmoozing with customers and watching for shoplifters while Rachel and Joseph stacked crates of soda and other supplies in the back. With his bad heart Naftali was too weak to lift anything. He did all the banking, in part because it would have been unseemly for his wife to handle the money.

Over time Naftali started drinking more and more. He took out his frustrations on Joseph, sometimes hitting him for his lack of reverence for Judaism. Once, after he found Joseph eating spareribs in a nearby Chinese restaurant, he almost knocked him out. The boy was only ten, but Naftali hit him hard. Rachel intervened, standing between them and protecting her son from further violence.

Joseph hated his father, and he still feels guilty that he was almost unmoved when Naftali dropped dead from a massive heart attack barely a month before Joseph's bar mitzvah. But today, with everything on the line, he misses him for the first time.

December 2, 1961. East Side Hebrew Institute, Lower East Side.
Rachel Helfgot sits with Pauline, her daughter, partially obscured by a screen in the auditorium of the Jewish religious school. Her jaw is firmly set and she stares blankly ahead, knowing the women who sit with her during services are watching to see if she will cry. She can barely endure the injustice of Joseph reading from the Torah without his father standing with him on the bimah. How could her stupid husband drop dead just before the bar mitzvah of their only son?

She holds her prayer book a little higher, pretending to read the Hebrew letters while she mumbles the prayers by rote. Rachel, who grew up poor in a Polish shtetl, never learned to read. When she was still a little girl she watched one evening as her mother prepared to light candles for Rosh Hashanah. She lifted the glass from the oil lamp and lit a small wad of paper from the wick. As she set the glass back into the lamp and turned to recite the blessing, she burned her finger and yelped in pain, knocking over the lamp with her elbow.

Rachel watched as oil spilled down her mother's dress and immediately caught fire. The flames engulfed her mother and Rachel ran from the house screaming for help. But all the men had left for the evening prayers, and there was nothing the girl could do. Rachel's little brother followed, his hair aflame. The house was consumed.

It took three horrible days for her mother to die, screaming in pain the whole time. Her brother survived, but he was badly scarred. The sounds of her mother's screams still haunt Rachel, more than all the other tragedies she has endured. And she has endured a great deal.

After her mother's death Rachel was sent to her father's cousin in Warsaw, a widower with a small infant. As she grew older he began to look at her in a way that made her uneasy. She fled the tiny apartment and lived off the streets, selling candy and flowers during the day and returning at night to the Jewish section, where a kind family gave her refuge.

She married the first young man who came along, and they had two beautiful children, a boy and a girl. They were desperately poor, but Rachel made extra money selling items she pilfered from expensive shops in the heart of Warsaw, returning to the Jewish district at night to sell her bounty. She hired a woman to look after her toddlers during the day. In the summer they would go on picnics at the lake near the palace, with fruit and bread and cheese that Rachel carried in a potato sack.

But the young family's tranquility was short-lived. Germany invaded Poland in 1939 and marched toward Warsaw, quickly surrounding the city. One night Rachel and her husband awoke to the sound of airplanes flying low. Sirens began to wail. An explosion rocked their building as the bombs started falling and continued through the night. The Second World War had erupted on their doorstep.

The Nazis entered the city a few weeks later. Soon the Jews were no longer allowed to leave the northern quadrant of Warsaw except during certain hours, and Rachel was forced to wear a white armband imprinted with a blue Star of David. Food was rationed. She tried to keep her children's stomachs full by giving them carrots. They ate so many that her daughter's skin began to turn orange.

Men were rounded up to work in German labor camps. Her husband was taken, and Rachel never saw him again. A brick wall was built around the Jewish quarter, the ghetto. When it was finished no one could come or go without permission. Rachel's radio was taken, and her son's tricycle. They told her it was for Germany's war effort.

One day a man came and asked for all the knives. He let Rachel keep one dinner knife. She complained that she couldn't cut anything with a dinner knife, but he turned away and began to walk out the door. She continued to rant until he finally turned around. She thought he was going to give her back one of the knives, but he walked up to her and slapped her hard across the face.

While they slowly starved, Rachel heard stories about Jewish children being taken from their homes in the night and sent to refugee camps, supposedly for their safety. She also heard rumors that children were being killed in those camps. Rachel didn't believe these stories, but she was afraid for her little ones, who were seven and five. A nurse came into the ghetto twice a week to check people for typhus. She was known to smuggle small children out in her medical bag, but there were thousands of children and just one woman with one bag. Rachel begged the nurse to take her tiny five-year-old daughter. The nurse said she would try, but there were so many children that Rachel shouldn't expect it.

Late one night soldiers with German shepherds broke down Rachel's door. They swarmed the apartment, hurrying the children into their coats while they shrieked in terror at the barking dogs on short leashes. "Mameh, don't let them take me!" her son cried out. "I don't want to go!"

"We're taking you where it's safe," one of the men said with a gruff German accent.

"No, no!" Rachel screamed. She flew at the soldier, trying to gouge out his eyes. He lowered the butt of his rifle over her head. She awoke the next morning, the sun streaming into the window. A small pool of blood had hardened on the floor next to where she had fallen. When she tried to stand, her right leg gave way. A chunk of flesh was missing from her calf, the dog bite already filling with pus.

She ran limping from the building, screaming that her children had been taken, but the street was full of crying parents. Later that afternoon a man who had slipped out through a dry canal returned with the news that a train had been seen leaving Warsaw earlier that day, filled with Jewish children. Rachel limped back to her apartment and poured her own urine on the dog bite, for there were no medical supplies anymore. She waited until the Nazis finally came for her.

December 2, 1961. East Side Hebrew Institute.
"Ma, stand up." Pauline nudges her mother.

Rachel is thinking about the deposit she gave to Little Hungary Restaurant over on East Houston several months before. They kept part of it, although it wasn't her fault that Naftali dropped dead and she can't make a party for her son. *Gonifs!* In mourning, they will make do with a modest lunch downstairs, after the service, with bagels and cream cheese. The Bremers are taking Rachel and her children out to dinner tonight, because there won't be a party. Mrs. Bremer too is an Auschwitz survivor. Maybe she can get Joseph a summer job at her cousin's place in the Catskills.

Rachel watches her son on the bimah with the other men. As though a thousand knives have not already shredded her heart, she has to endure Joseph's bar mitzvah as a widow. But her eyes are dry. She hasn't cried in years. She squares her shoulders, reciting the prayers she knows by heart, but is unable to conjure up any communion with her Creator. What has she done in her life to deserve this added insult? It's not that she loved her dead husband, who was often terrible to Joseph, especially around matters of religious observance. Sometimes Rachel spots her son sitting alone on a park bench, pulling spareribs out of a small red and white bag. She leaves him alone. It's a small offense, and she is no longer sure God is watching. Or that God even cares.

Spring 1970. Manhattan Draft Board.
Joseph is wearing a tie-dyed T-shirt with the initials SDS emblazoned on the back. Two years ago Students for a Democratic Society shut down a number of college campuses to protest the war. Although the provocative shirt adds to the drama, the SDS can't help him now. Four months ago he sat on his mother's couch in Brooklyn, watching the national draft lottery on television. His birthday came up 123rd out of 366, only a third of the way down. He knew this meant Vietnam. He told himself that he hadn't been born of Holocaust survivors to go halfway around the world and die in a swamp.

The words of the draft counselor rush back to him. "Whatever you do, don't sit. Stand. Pace. Don't get into line. Push your way to the front. Be as rude and as different as possible." For Joseph, who grew up in a tough neighborhood, these instructions aren't hard to follow.

He pushes his way to the front and rips a clipboard from the hands of the man who is giving them out. Leaning against a wall, he checks "yes" to every question on the medical form: bedwetting, night terrors, drug addiction, even attempted suicide.

He throws the clipboard on the table and stands around waiting for the medical exam. The men are told to strip to their under-wear. "You won't be wearing underwear," the counselor had said.

As he stands in his altogether holding his clothes, the oth-ers give him a wide berth. He is beginning to attract attention, which is good. An MP walks over to him. "You think you're being funny? Didn't your mother teach you to how to dress?"

Joseph spews out a tirade in Yiddish, spitting words at the man until the MP rips the pants out of Joseph's hands and throws them in his face. "Put these back on, now!"

Joseph lets the pants fall to the ground and proceeds to uri-nate on them. The MP grabs him and pushes him down the hall to the psychiatric service, where he shoves Joseph behind a curtain. A few minutes later a middle-aged doctor with glasses pulls the curtain aside. Joseph, buck naked, lunges at the man, ripping his white coat pocket, scattering his pens, and knocking his glasses into the air. When the doctor tries to retreat and get help, Joseph runs after him, grabbing at his stethoscope.

"Get this f—ing kid out of here!" the psychiatrist shouts. The MPs throw Joseph out of the building.

Several months later he receives his 4F status: unfit for mili-tary service.

chapter twenty-two

Visitors have been arriving in a steady stream all weekend.

"Josie," Susan says, "you made something? Really, you didn't have to." She takes a large Tupperware bowl from the woman's hands. Her eyebrows stiffen as she opens the plastic lid and peeks at the contents. *What the hell is this?* "Let's go in the living room. Would you like a glass of wine?"

It's only four o'clock. They still have the evening prayers tonight at seven. Susan isn't sure she can stay upright that long. The phone rings for what seems like the millionth time.

"Sue, someone named Sharon is on the phone." Sharon is Claude's wife, and Susan is reminded of the merguez sausage. She feels tears forming. *What is it about the damn sausage that makes me so weepy?*

She closes the bedroom door. "I am so sorry I wasn't there," Sharon says. "Claude just got home from the airport."

Sharon has been stuck in the last round of financing for a company she recently started. Susan cradles the phone to her ear, looking at the picture of the two smiling couples on the bedside table. Sharon was very pregnant with Daniel. "How is Daniel?" she asks.

"Can you believe his bar mitzvah is next year?"

"No, I can't. I went to his bris only a week ago."

"Listen, Sue, Claude told me about Joseph's face."

"Mm."

"I can't believe you did that. It's amazing."

"You can't tell anybody."

"I won't, but you know it will probably get out. Things like this always do. So many people knew Joseph, and it's a big story, so don't be surprised."

"Okay, I promise if it gets out, I'll let you have the story, okay?" Sharon is an entertainment reporter.

"Sue, how are you doing, the truth?"

"The truth? Really crappy."

They sit quietly for a few moments. "I'm really sorry." Sharon is crying.

Someone calls up the stairs. "Sue, some people are here from the hospital."

"I'll be right down."

"Sue, Joseph suffered a long time. You both went through hell. Don't forget that. He's not suffering anymore. When I saw him last summer with the feeding tube and that purple thing sticking out of his throat . . ." Her voice trails off.

"I know."

Susan heads down the stairs.

Four years earlier. Boston.

Susan stands under a hot shower after spending the day at the Brigham. Things are looking grim for her mother, who has been in the hospital for several weeks. Everything seems to be conspiring against the worn-out body her mother inhabits. But hanging around the ICU is oddly soothing, as the whir and click of medi-cal equipment and the low hum of voices and beepers combine in a soft white noise. No kids, no doorbells, no cell phones.

Being there takes up a lot of her time, and Susan must squeeze the rest of her day into a few hours. She's grateful that Chanukah is already over and that it didn't overlap with Christmas this year. She wonders if her mother will make it to Christmas. She forces her mind to bend toward the positive.

Stepping into the house a few hours ago Susan was greeted by the familiar pile of shoes and backpacks from half the kids in the neighborhood. The shoes always strike her as far too large for teenagers. The dog was on the table, shredding a bag of potato chips. "Freckles! Get down!"

"Mom?" Ben called out. "There's something in my room. *Shit! It's a bat!*"

"What? I can't hear you. And don't say *shit*." She set down a bag of groceries.

"Jesus, Mom. I said *there's a bat in my room.*"

"It may have rabies. Get out and shut the door."

Great. She rummaged around the front hall closet and pulled out her old tennis racket. She had heard that bats can't detect tennis rackets. Would her paltry tennis skills be enough to go up against one measly little bat? If it even was a bat.

Susan carefully cracked the door. It was a bat. She waited until it circled around the room again. On its third pass, she reached out and *thwack!* The bat landed hard against the opposite wall and dropped dead to the floor. It was so tiny. She picked it up with tissue paper and flushed it down the toilet. A mother's work is never done.

"Ben, you can come in now."

"Thanks, Mom." He leaned over and kissed her on the head.

"Now do your homework."

• • •

A shower feels like heaven after a day like today. *What now?* Ben is shouting something through the bathroom door. "What, Ben? I can't hear you."

"I think there's another bat in my room," he yells.

"Get your father," she shouts back, lathering shampoo on her head. Wait, maybe that's not such a great idea. Joseph has a defibrillator in his chest. She turns off the water. "Hold on, I'm coming." She wraps herself in a robe and runs into Ben's room, soap dripping from her hair.

"It's up there." Ben points to a shelf high above his desk. Joseph has one foot on a swivel chair.

"Joseph! Are you crazy? Don't get on that chair. You'll fall and kill yourself." She lugs a large ottoman across the room.

"Susan, put that down, it's too heavy for you." Joseph steps on the ottoman and looks around, pulling books from the shelf. "I don't see anything."

"I'm getting back in the shower. Ben, put the ottoman away." Joseph isn't allowed to carry anything heavier than five pounds.

Susan has finished her shower and grabs a towel. She hears Ben call, "Daaad, I hear it again."

A few seconds later she hears a loud crash.

"Mom! Dad fell!"

She slips running back from the bathroom. Joseph is sprawled out on what's left of Ben's desk. His chest has landed on Ben's heavy, boxy computer monitor, and his arms and legs dangle in midair on either side. The swivel chair is on its side, one wheel still spinning. Books and papers litter the floor. Ben is teary eyed.

"I *told* you not to get on that chair!"

Joseph follows her, limping into their bedroom, and gingerly lies down on the bed. "I'm in pain."

"Good!" She tugs on her nightgown and leaves to help Ben clean up the mess.

In the middle of the night, after Joseph has been tossing and turning for hours, Susan hears him reach for the remote. "If you turn on the TV, I will divorce you."

"I thought you were asleep."

"I am *trying* to be asleep."

"What is the dog doing in my bed?" Freckles cautiously eyes his master. The sweet cocker spaniel was picked out by Joseph in a moment of weakness. He has never been able to say no to his children. When Joseph is away the dog sleeps with Susan. When Joseph is home, he tries.

"Please, can't he stay just this once?"

"If I can't watch TV, he can't stay in my bed."

"Freckles, get off the bed," Susan says gently. The dog looks at her and wags his tail.

Joseph lowers his voice. "Off." Freckles jumps off the bed and prances into Jacob's room.

Two hours later Susan is awakened by the sound of the kitchen door closing. "Joseph?" No response. The clock reads a little past 4 a.m. She heads downstairs. The kitchen is cold and smells of winter. Opening the outer door, she sees Joseph limping around the corner of the house, toward a waiting taxi. She runs after him in her bare feet, her nightgown no match for the bitter wind. "What are you doing?"

"I'm going to the hospital."

"I'll take you."

"Go back to bed."

"Don't be ridiculous."

"Go back inside. It's cold."

"Joseph, please don't—" But he is already in the cab, which is speeding away. She runs inside and calls his BlackBerry. "I'm coming to the hospital."

"No, stay with the kids. I think I might have broken something. I'm in a lot of pain."

"Are you sure?"

"About the pain?" There is a pause. "I'm sorry. I shouldn't have stood on the chair. It was stupid." He sounds like a penitent teenager.

"It's all right."

"I love you, Suze."

"I'm coming." She is worried.

"Take your time. Get the kids off to school. Nothing moves fast in a hospital."

She shakes Ben awake.

"Is Dad okay?"

"He's fine. They're just checking him out to be sure. I'm going over there now. Make sure Jacob gets to school."

"Okay."

In the Emergency Room Joseph is connected to a morphine drip. He has two broken vertebrae and three broken ribs.

Back home he spends New Year's Day on the couch, watching the Rose Bowl game and reading movie scripts, popping painkillers as though they are candy. They seem to be having little effect. "Do I remind you of House?" he asks Ben. Gregory House is a fictional TV doctor with a drug habit.

"Dad, give me a break."

Six weeks later. Helfgot residence.

Joseph hasn't slept in days. He is still in intense pain from the fall and is very cranky. Susan is heading over to the Brigham to visit her mother, who has recently been admitted for the third time in six months. She too is cranky. Dan the plumber has just arrived to fix a leaky shower, and she feels better knowing that someone is in the house with Joseph while she runs over to the hospital.

Joseph looks like hell. Too much pain medication and too

little sleep is a bad mix. And her mother is back in the ICU, wearing a hockey mask that pushes air in her face, keeping her alive. She looks like Jason in *Friday the 13th*. Susan looks in the closet for something to wear.

Joseph calls out from the bed. "Suze?"

"What?"

"Why is my computer floating in the middle of the air?"

"Whaaaaat?" She turns around. The laptop is sitting on the bed.

"There." He points to a spot in the middle of the room. "I don't feel well," he says in a weak voice.

"That's it. We're going to the hospital. Dan!" Rubbing his dirty hands on his pants, the plumber walks into the bedroom. "Help me get Joseph into the car."

"Susan?" Joseph calls out into the air. "I can't see you."

Oh my God! She considers calling an ambulance, but she can get him there faster herself. They half-carry him downstairs and push him into the back of the car. Susan races over to the Brigham, two miles away. As she pulls up to the Emergency entrance, a cop comes over. "Ma'am, you can't park here."

"Heart attack!" she shouts. Experience has taught her that these are the magic words. A wheelchair appears out of nowhere and someone takes her keys.

"Please page Dr. Stevenson," she calls out to the receptionist. The woman sends them straight to triage. The Helfgots have been here more than a few times.

"I can't get a blood pressure."

"Where is Dr. Stevenson?"

Soon Carol Flavell arrives. A nurse practitioner with the heart failure group, she has been following Joseph closely for years. Not her most compliant patient, he once apologized for his bad behavior by sending her an enormous fruit basket. The cellophane began to rip as she lugged it down the Pike, and a cascade of grapefruits and tangerines began rolling in various directions. It was quite a sight, watching senior physicians chasing after the runaway fruit.

"What's happening?" she asks Susan.

"They can't get a b.p."

"Take it again," Carol tells the attendant.

"Fifty over ten."

"Let's get him into the crash bed. Susan, did you bring his interrogator?" In addition to his defibrillator-pacemaker, Joseph has a new, experimental device in his chest that records fluid pressures in the heart. The device is read by an "interrogator," an instrument they keep in a drawer next to the bed.

"I just wanted to get him here as fast as I could," Susan says apologetically.

"It's okay. Could you run home and get it? It would really help us to have it."

"I'll be right back."

On her way out the kitchen door, interrogator in hand, Susan hears her cell phone ring.

"It's Carol. We're losing him."

"Oh my God."

"Come fast. Can you get someone to drive you? Sue, are you there?"

"Yeah." She is opening the car door.

"Get someone to bring you."

"Okay." She throws the phone on the floor and reverses the car out of her driveway, narrowly missing a garbage truck driver who has just hopped out of his cab.

A young social worker is standing at the emergency room entrance and escorts Susan back to the crash room. Curtains enshroud the space, and the two women stand anxiously on the other side, listening for clues. People suddenly start running toward the curtain as Susan hears the words "Code Blue, Code

Blue" on the intercom. *Oh my God!* A young doctor runs up. It's her mother's cardiologist.

"What is your mother doing down here?" he asks, confusion on his face.

"It's my hus—" Susan starts to say, but he has already pulled the curtain closed.

"Three thirty-seven." Lynne Stevenson finally pulls back the curtain. Susan feels her knees buckle. The social worker steadies her.

"Three thirty-seven what?" she asks.

"Minutes he was dead—this time. The first time he was out about a minute and a half."

"That's when I called you," Carol says.

Her mother's cardiologist says, "You aren't having a very good week, are you?" She shakes her head and looks at Joseph, who has a tube down his throat.

As they prepare to take him up to the ICU, Dr. Stevenson whispers, "I got his body back, but I don't know about his mind. He was out a very long time."

A few days later. Intensive Care Unit.
A woman comes to the door of Joseph's room where Susan is sitting. "Mrs. Helfgot, I'm Dr. Kim. I'm a psychiatrist working with the Trauma Unit." They shake hands.

Joseph is sitting up in bed, watching TV. "Mr. Helfgot, you had a little trouble last night?"

"Who are you?"

"I'm Dr. Christine Kim."

"Susan, can I talk to you?"

"Sure, honey."

"Without *her*."

Dr. Kim moves out of earshot. Joseph pulls his wife down to the bed and whispers in her ear. "Last night they turned everything into a reality TV show. About doctors. It's a Japanese show. All the doctors are Japanese. They come in at night. That teleprompter," he says, pointing to the monitor on the wall, "it's in Japanese. See?"

Susan looks up at the monitor. Joseph's heartbeat is being graphed in a green line, his EKG showing irregular shallow heartbeats. She nods her head, and he pulls her even closer. "The people who come at night to do the show? I think *she* is one of them." He gently nods in Dr. Kim's direction.

Yesterday Joseph convinced the nurse on duty to call the house. Susan was down in the hospital coffee shop. Jacob picked up the phone. "Dad, is that you?"

"Jacoby, when are you coming? This boat is really cool."

"Dad?"

"I told them the ship can't leave on the cruise until you and Ben get here too. I keep waiting for you. I'm really mad at you for not coming."

Ben grabbed the phone from his younger brother. "Dad, it's Ben. We'll call you back, okay?" Jacob had started to cry.

Dr. Kim steps outside with Susan. "Sometimes it just takes time. Your husband is on a lot of medication."

"But he'll eventually be all right, won't he?"

"It takes time, but usually, yes."

Usually?

A week later Dr. Kim steps into Joseph's room. He is sitting on the edge of his bed, papers scattered everywhere, banging away on his laptop like a madman. As he shifts his weight to look up, papers slide off the bed. "Do you go to the movies?" he asks her.

"Sometimes. How are you feeling today, Mr. Helfgot? I heard you had a bit of trouble again last night."

"Yeah, I guess I did. I thought they were making a movie here, and the ICU was a casino." He gets up and walks around the room. One of his heart leads pops off, triggering an alarm. His nurse comes in and admonishes him. "Stay in bed."

"It was so real. But I remember the whole thing exactly."

"So the next time this happens, when something doesn't seem to make any sense, you can remember that you've had an experience like it before. Can you try to do that, Mr. Helfgot?"

He gets up off the bed and walks to the door of his room. All the leads snap off. He points to the row of monitors next to the nurses' station. "I thought those were teleprompters. They were so *real*. Can you believe it?" He returns to the bed and starts re-attaching the leads. His nurse rushes in. "Joseph, you're killing me. Put your hands down. They have to go a certain way."

"Mr. Helfgot, often when people see things that aren't real, they see something that they can relate to, something they're already familiar with—in your case, the movies."

"But it was *so* real. So do *you* know when I'm going to get out of here? No one seems to know."

"That's up to Dr. Stevenson. How is your pain?"

"Lousy. I need you to write something to help me sleep."

"We'll work on that."

"I thought Kevin, my nurse yesterday, was Fat Gerry, this drug dealer I used to know. Kevin doesn't even look Italian. He's not even fat."

"I'll see you tomorrow." Dr. Kim scribbles something in her notes. "Remember, you're here at the Brigham. We're a hospital. We don't make movies here." She smiles at him and points to a sign Susan has put up on the wall. YOU ARE AT BRIGHAM AND WOMEN'S HOSPITAL. "You're doing much better, Mr. Helfgot. See you tomorrow."

• • •

Monday, April 13, 2009. Trauma and Burn Unit.
Dr. Pomahac has just finished checking James Maki's face. The microvascular connections seem to be holding, and the swelling is abating. So far there has been no rejection.

"When do you think I can look at myself?"

"Let's see what Dr. Kim thinks. Do you feel ready?"

"Definitely. Let's do it."

"Do you think he's ready?" Dr. Pomahac asks Dr. Kim later that day.

"Mr. Maki says he's ready. I think he is."

"Then let's set up a time. Peter told me the crew wants to film it."

"But it's such a private moment," she says. "No one knows how a person may react in that situation."

When Dr. Pomahac began to review potential candidates for facial allograft surgery, he needed to satisfy two important criteria. The first was this: of all the patients in his care, who would benefit the most from such an extreme surgical procedure? Everyone agreed, including the surgeons at the meeting in Brussels, that Pomahac's most seriously disfigured patient was James Maki. Another dozen surgeries would do little to improve his ability to speak, swallow, or breathe, basic functions that the rest of humanity takes for granted.

The second criterion was whether Jim Maki was physically and emotionally strong enough to endure the marathon surgery and its long aftermath. Dr. Kim spent months meeting with him and concluded that he wanted the surgery, understood what it entailed, and would be able to cope with his new identity.

Jim has been desperate to learn what life with a new face might hold. Four years of sobriety in isolation, enduring surgery after surgery from the subway fall that should have killed him, left him a changed man. He knew he was on God's time now and wanted to make good on the years he had left. Without a face, that would be impossible.

His doctors believed he deserved a chance. And so James Perry Maki became the first person listed for a facial transplant at the Brigham.

Dr. Kim is appropriately concerned for her patient. It can't be easy to wake up with someone else's face. Whose nose am I wearing? Whose teeth are in my mouth? Mr. Maki has asked more than a few times who died to give him this face. It is good that he registers the enormity of this gift, that someone had to die in order for him to be whole. It will take time before he merges with his new face.

Dr. Kim understands what he's going through. She too still sees another man's face when she visits with Maki. And like her patient, she wonders about that other man and what kind of life he might have led.

chapter twenty-three

Monday, April 13, 2009, late morning.
The offices of the Boston Globe.

friday's face transplant announcement was big news. After her interview with Dr. Pomahac, Kay Lazar spoke with four other sources and sifted through the literature on face transplants, including an article on Dr. Pomahac's three-year quest to perform the operation at the Brigham. Somehow she managed to put it all together before her deadline. A quick Friday afternoon posting on White Coat Notes ran ahead of her full story in Saturday's paper. She has already moved on to her next assignment.

The blogger for White Coat Notes is reading e-mail comments from readers, including quite a few about the transplant. She spends a lot of her time screening the postings that stream into the newspaper's website, filtering them for believability and a semblance of decorum. Today, as she skims the postings, she reads one that stops her cold.

"Kay, look at this." A man has written that during a Torah study session at his Brookline synagogue, the rabbi cited the face

transplant as an example of the ultimate mitzvah, an act of kind-
ness that can never be directly repaid. And apparently he mentioned that the donor had been a member of the congregation.

Really?

Kay stares at the words for a few seconds. So the unknown
donor may not be so anonymous after all. She quickly types a
response to the person who submitted the comment: "We
would very much like to share with our readers, in a sensitive
and thoughtful way, this exceptional person's story of generosity.
Could you contact me?"

"I don't know the man," the correspondent responds. "The
rabbi didn't mention him by name. Let me check with him and get
back to you. I can tell you he is from Brookline and in the entertainment business." The day drags on, but Kay hears nothing further.

She starts scanning the local death notices. Nothing rings a
bell, so she checks the websites of *The Hollywood Reporter* and
Variety. In *Variety* she finds an obituary that begins, "Joseph
Helfgot, the sociologist who founded the media market research
firm Marketcast, died from complications of a heart transplant
Wednesday in Boston. He was sixty."

In Boston—died from a heart transplant!

Kay's mind races. She replays the press conference in her
mind, trying to recall exactly what was said about the donor. She
pulls up the statement released by his family. "To go from being
a recipient family to a donor family so suddenly has given us the
opportunity to fully understand the power of organ transplants
to give and transform lives." *The donor was a transplant recipient.*
Kay pulls up her newspaper's obituary for Joseph Helfgot.

"Brookline resident Joseph Helfgot is survived by his wife,
Susan, and three sons and a daughter. He was sixty years of age."
Brookline. The Torah study session was in Brookline. And there
are condolence notices from Los Angeles posted on the funeral
home's website. The donor has to be Joseph Helfgot!

Kay and her associates mull over what to do next. "Just drive over to the house," someone suggests. It takes them only a moment to find the family's address.

"I don't know," Kay says. "Wouldn't that be a complete shock to his widow?" Kay isn't a hard-nosed reporter, and she hates interviewing grief-stricken families. She has chiseled her career around an aversion to situations like this. Now she has to make a tough choice. "Is the family still sitting shiva?" she wonders aloud. They look at the obit again. Visiting hours were Friday through Sunday.

A few hours go by. The man from the temple still hasn't responded. "Kay, just get in the car and go." But she decides to wait out the day.

Later that day. Trauma and Burn Unit.

Dr. Kim's concern for her patient is evident on her face. Maki's hospital room is crowded with people. She hopes he won't be overwhelmed by so much attention for what should be such an intimate moment. Dr. Pomahac stands ready to watch his patient's reaction to his handiwork. Maki's nurse checks his face. She attends to some minor detail and runs a comb through his dark hair so he will look his best.

Cameras whir quietly in the background. Although Dr. Kim's concern is well intentioned, it turns out to be unnecessary. Jim Maki is so excited that later he will be surprised to learn that a TV crew was in the room. What he will always remember is what awaited him when he finally saw his new face.

Dr. Kim shares a knowing look with the nurse practitioner in charge of Maki's care. Maki is only four days out from surgery, and normally they would never allow so many people in a patient's room. They are crowding around the bed, and Maki's nurse will be glad when the hubbub dies down and she can get back to taking care of him.

"Are you ready?" asks Dr. Kim.

"Yep."

"You're not nervous?" Dr. Pomahac asks.

"No, I'm excited. I want to see my face."

The room falls silent. After a moment someone asks, "Who has the mirror?"

People look at Dr. Kim, and then at the nurse, and finally at Dr. Pomahac, as though a surgeon might have a mirror at the ready. A few chuckles finally explode into full-blown laughter. No one has thought to bring a mirror. A nurse runs from the room, frantically making her way around the floor. But there are no mirrors in the Trauma and Burn Unit. Some patients need time to get used to their deformities. Others never do.

She bursts back into the room. "Here," she says, and hands Dr. Kim a tiny makeup mirror.

"Oh, we can't give Mr. Maki that," says Dr. Kim. "It's too small. Can we find something a little larger?" The nurse leaves again and the room falls silent once more. The climax they have been waiting for still eludes them, and for the most mundane of reasons. But they all see the humor in this moment: in a building filled with some of the most sophisticated and expensive equipment ever devised, they are waiting for someone to bring in one of the oldest, simplest tools in recorded history.

Finally the nurse returns with a large square mirror she has pilfered from another floor. She hands it to Dr. Kim. "How's this?" she asks.

"Perfect. Are you all set?" asks Dr. Kim.

"Yes."

She holds the mirror for him as he takes his first long look at himself, moving his head carefully left and then right to inspect his cheeks, and then raising his chin to see under his neck. Looking at Pomahac, he says, "The guy who orchestrated this did a good job."

Maki's facial nerves, which were microsurgically joined to Helfgot's, have not yet healed, so he feels no sensation of any kind, not even pain. He touches the new skin, his fingertips lightly resting on something soft and slightly rubbery. Although the mirror suggests that this new material somehow belongs to him, this isn't *his* face, he realizes as he studies the image. The nose belonged to another man. These teeth and upper lip used to live in someone else's mouth, drinking his water and speaking his words.

The monstrosity that was his previous face is gone. He has returned to life as a full human being. He is humbled. He is shocked. And he is grateful.

There are long, red suture lines running from the corners of his mouth down along the jowls and back up the sides of his face over the cheeks. The lines join perfectly at a common point in the center of the bridge of his new nose. His whole face is still swollen from the surgery, the puffiness exacerbated by large doses of antirejection drugs. But to a man who has lived with a hole in the center of his face for four long years, these imperfections are almost invisible.

He looks up at Dr. Pomahac and says, "My new face looks just like my old one."

It's quite a stretch, but nobody is going to argue.

Bo Pomahac returns Maki's gaze as one man to another, rather than as doctor to patient. Maki's features are swollen and scar ridden, but otherwise normal. The faith that this damaged and fragile man had placed in him, and the risks they had all taken—the naysayers were wrong. This *was* the right thing to do.

And there are plenty of others out there, Pomahac knows, including soldiers with terrible injuries and people with severe burns or tumor disfigurements, who will take hope from the news of this groundbreaking operation. Each one of them, like James Maki, is desperately hoping to feel human again.

chapter twenty-four

Tuesday, April 14, 2009. Helfgot residence.

"S ue, there's a call for you from the *Boston Globe*."

Susan is pulling clothes out of the dryer and putting them into Jacob's outstretched arms. A week ago today she was at the hospital, where Dr. Rawn was having trouble looking her in the eye. *Has it been a week already?* Earlier this morning Jacob tactfully reminded his mother that he was out of clean underwear and that maybe she should think about doing some laundry.

"I'm not ordering the *Globe*. Tell them someone died or they'll keep calling back."

"She says she's a reporter."

A reporter? Why on earth is she calling? A local paper ran a story on Joseph when we first moved back here from California. Or maybe it's for an obituary. "Can you tell her it's not a good time?"

"She says it's important."

"I'm coming." Susan picks up the phone.

"Mrs. Helfgot, this is Kay Lazar with the *Boston Globe*."

"Yes?"

"I want to tell you that I am very sorry for your loss."

"Thank you."

"I would like to speak with you. It's about your husband."

Something isn't right. Her voice—I can hear it in her voice. "What about my husband?"

But Kay is too experienced to say any more on the phone. And she hates this kind of call. "May I come and see you?"

This woman sounds mortified. Something is wrong.

"I guess this afternoon would be fine," Susan says.

As I stand with the phone to my ear, I watch the people in the kitchen moving all around me. They are in one world, the one I used to live in. This reporter is taking me into another place. I think she knows.

Emily's and Jonathan's luggage is on the kitchen floor. I can't bear that they're actually going home. While we are all together I feel safe. This woman is coming over and something is about to happen. And they won't be here with me.

She picks up the phone. "Sharon?"

"Sue, how are you doing?"

"Listen, I think they know. What am I going to do?"

"Who knows?"

"The *Globe.*"

"What do you mean you *think* they know?"

"A reporter called. She's coming over to see me about Joseph. Her voice, Sharon, she knows something. What should I do?"

"If she already knows, there's nothing you can do. You could ask her not to break it, but I don't think she would sit on a story like this. It's too big."

"I know."

"Sue, it's your story. You should do whatever you think is right."

"If she knows, I'll tell her to hold it so you can run it first, okay?" Sharon runs a website called TheWrap.

"Thanks. Try to make her do all the talking. She may not know as much as you think."

Kay Lazar is annoyed with herself as she rings the doorbell. How could she show up late for this interview? She called Susan from the road, but this could be a difficult conversation and she doesn't like starting off on the wrong foot.

A petite woman whose age Kay can't quite estimate opens the door with a smile.

"Hi." Susan extends her hand in greeting. She studies the reporter's face and is met with empathetic eyes. Kay's sympathy for the new widow is genuine, but Susan detects something else. Discomfort, certainly. Anxiety, perhaps. *I think she knows Joseph was the donor.*

Susan has been hoping that Kay was coming for some other reason. Unlike Joseph, who enjoyed public attention and loved mixing with celebrities, she has long had an aversion to fame. In the ninth grade she watched as her classmate Chris Evert became an overnight tennis sensation. The paparazzi showed up at school and the poor girl couldn't go anywhere. It was, Susan felt later, a little like having leprosy in reverse. Even now, as she guesses that the jig is up, she still hopes to remain anonymous.

Kay follows her into the living room. In a Jewish house of mourning it is customary to cover the mirrors, as vanity is temporarily suspended by grief. In this house the mirrors are crowded with photographs, thickly layered, one on top of another and strung together with cellophane tape. Frozen memories of a man who died too soon.

"Mrs. Helfgot," Kay starts.

"Please, call me Susan." They sit a moment too long. Finally, "It's all right, Kay. I think I know why you're here." Lazar nods a yes.

•••

Shortly after Kay leaves the phone rings. "Mrs. Helfgot? I'm Peter Brown from the Public Affairs Office at Brigham and Women's Hospital. I want to express my sympathy for your loss."

"Thank you."

"Kay Lazar from the *Boston Globe* called me just now. She told me that you two discussed your husband's gift. She was wondering if you would mind her speaking with a few people here. Dr. Stevenson is willing to share her thoughts if that's all right with you."

"I . . . sure, that's fine." After the call Susan walks outside into the yard. She sits in the garden that is still struggling to fight off the long New England winter. A friend comes outside to check on her.

"Sue, are you okay? Who was that on the phone?"

"The hospital wanted my permission for the *Globe* to talk to Joseph's doctors. It hit me during the conversation that the man who got his face is going to know who we are."

"Is that bad?"

"It could be. I suddenly remembered that he has the right to decide whether he wants to know about the donor. What if he doesn't? There's a whole system in place, and I forgot all about it. Donor families and recipients can write letters to each other, but without giving their names. The organ bank passes the notes along, but if either side doesn't respond, everyone stays anonymous. When I spoke to the reporter I wasn't even thinking about that. Maybe I should have called Esther from the organ bank before I spoke to the *Globe*. I've made a big mistake. What am I going to do?"

"Sue, calm down. You have been through so much. Don't beat yourself up. It's not as if you do this every day." She pauses, and then adds, "Thank God."

"I feel terrible."

"Don't. You did a wonderful thing, giving that man Joseph's face. I bet he wants to know whose face he's wearing. I'd want to know. Wouldn't you?"

"I guess."

"Stop worrying about it. You have enough to deal with."

"I have a bad feeling this whole thing is going to blow up in my face. Why are you looking at me like that?"

"Did you hear what you said? Blow up in your face?"

"Oh, very funny."

"Sue, did *you* call the *Globe*? No, they called you. Take it one step at a time. It'll all work out."

Later that day. Office of Public Affairs,
Brigham and Women's Hospital.
Peter Brown is with Dr. Lynne Stevenson. They have Kay Lazar on speakerphone.

"Kay, I have Joseph's cardiologist here with me. Mrs. Helfgot gave us permission to speak with you."

Kay asks Stevenson, "Do you remember when Joseph Helfgot's son had his bar mitzvah last year? His wife told me that's all he lived for."

"He always talked about staying alive for that day. He was so extremely ill. We really weren't sure if he would make it with a trachea tube and the VAD." She pauses for a moment. "Joseph lent me a videotape of his mother describing her experiences during the Holocaust, and how she would sacrifice anything for her children, and her sorrow that Joseph's father had not lived to be at their son's bar mitzvah. That's what Joseph was living for, this passage for *his* son. He said the additional time was worth all the suffering he had endured.

"The last time I saw him, he told me that the few months

when he was able to leave the hospital for the bar mitzvah and his sixtieth birthday party were the richest of his life." She clears her throat. "He told me that even if there was nothing after that, it was all worth it." She stops speaking, unable to continue.

Wednesday, April 15, 2009, early morning. Boston.
Christine Kim relaxes in her favorite chair, checking her laptop to see whether anything new has been written about her patient. Although he remains anonymous at the moment, he is far from unknown. And whatever she reads, he will likely read someday.

She worries: How will he cope if his name is revealed? What if people think he isn't worthy of this gift when they find out about his past? Between heroin and jail, it's not a pretty story.

Mr. Maki has spent four long years recovering from his hellish fall onto the subway tracks. At first, he told Dr. Kim, he didn't even realize that his face had been destroyed. It took a long time for his mind to clear after the accident, which wasn't surprising. After being severely electrocuted he was on life support for several weeks. He was also withdrawing from the effects of the various illicit drugs he had been using while taking strong medication to stave off the pain of his injuries. His mind was foggy. He wasn't fully conscious the first few times Dr. Pomahac took him to the operating room.

After three or four months in the hospital he woke up one day fifty miles south of Boston in the severely disabled unit of the New Bedford Rehabilitation Hospital. During the two years he spent there his mind gradually began to clear. The first time he saw himself in the mirror, he couldn't believe the image staring back at him. Everything from his lower lip up to his eyes was gone. No upper teeth or mouth, not even a nose. All that was left was a hole so deep it was black inside. He was looking at a man without a face. It would be shocking enough on another person, but to see *yourself* like that?

The unit was mostly filled with people whose brain injuries had left them virtually mindless. But Jim Maki slowly climbed his way out of that facility, and then rotated through other institutions before moving to a veterans' home outside of Boston. During that time he had multiple surgeries back at the hospital. A trachea tube and a feeding tube were his constant companions.

He knew the accident should have killed him, and he spent four years as an invalid, most of it flat on his back, reflecting on that fact. Dr. Kim knows that patients in such dire situations typically experience one of two reactions: either complete despair or a Herculean resolve to find some meaning in the madness.

Mr. Maki was in the second group, and his optimistic outlook has amazed her. She met him well before anyone had even mentioned the possibility of a face transplant. But even then he was calm, taking each day as it came and never complaining about his situation. In one respect he was actually lucky. He was rushed to the Brigham the night he fell, straight into the hands of Dr. Bohdan Pomahac.

She clicks on Favorites and waits for "Boston Face Transplant" entries to pop up. She scans the first one. The *Boston Globe*: "Gift of a face a testament to donor's enduring values." What? The donor's family has spoken to a newspaper? Hadn't they asked to remain anonymous? She knows the donor was a Brigham patient who died in transplant surgery, but that is the extent of the information given to the press—and to her. Mr. Maki is eager to know who gave him this precious gift, this second chance, but she hasn't been able to help him. The ramifications begin to swirl in her brain as she clicks on the article.

The front page of the *Globe* leaps out at her, with a picture of a handsome, middle-aged couple. They are dressed casually, and the woman's arms are entwined in her husband's. Their heads touch as she leans into his shoulder. They are smiling, and they seem intimately connected to each other and completely at peace. They are obviously in love.

Dr. Kim's mouth turns dry. She cannot take her eyes away from the man's face as he stares back at her from the screen. She knows this face. It's Mr. Helfgot!

Frozen in her chair, she gazes at the picture. As she reads Kay Lazar's article, she cries.

7 a.m. Helfgot residence.
Susan's alarm goes off and she wakens with a start. She is sitting up in bed with the light on, a book half-open in her lap, her glasses on her face. She was up at four, thinking about the recipient and missing her husband. She finally dozed off while reading the same page for the second time, or maybe the third. She can't get the recipient out of her mind. Will he be upset when he finds out she has shared her story? She's been half-expecting the organ bank to call. Kay Lazar was in touch with the hospital. She must have spoken with the organ bank. Maybe they're angry with her for speaking to a reporter.

She reaches for her cell phone and shuts off the sound as it begins another high-pitched scream. Jacob must have been fooling with her ringtones again, because the alarm mimics a test of the Emergency Broadcast System. *God, get me through this day.*

While she waits for the coffee to brew, she pulls up the *Globe* online. The picture her friend Ellie took several years ago looks so different on a computer screen. It sits above the headline, and she and Joseph seem to be other people. But she likes the story Kay has written. It is intended to make a person weep, and it does.

Although it's barely seven, the kitchen doorbell rings. It must be Jacob's friend picking up his backpack. The kid leaves his stuff all over town; it's a wonder he finds his way home at night. Still in her pajamas, backpack in hand, she opens the kitchen door.

But instead of the boy with the missing backpack, she finds a man on her stoop with a microphone in his hand. Right behind him is a another man, holding a camera. The man with the mike looks at her expectantly and opens his mouth to speak, but she slams the door.

Freckles runs into the family room and jumps on top of the couch, barking wildly in front of the window. Susan follows him and cracks the shade just a bit. Several dark SUVs are parked along the curb with people inside. A few people stand on the sidewalk, including the man with the mike and his friend. He takes a long drag from a cigarette while he talks on a cell phone. *Holy crap. I didn't see this coming.*

Trauma and Burn Unit.
Nurse Lorrie MacDonald takes a practiced look at her patient's nose and gingerly places a tissue against his nostrils, using a few quick dabs to sop up a bit of nasal drip. In this case the dripping is a good sign, indicating that Maki's new nose is working. A week after the surgery Lorrie is still tentative in her ministrations. The last thing they need is for any of the tiny, newly attached vessels or nerves to tear. It's all a bit nerve-wracking, she thinks, and smiles to herself at the unspoken pun.

Jim is trying to say something. It is remarkable that a man who was unable to have a real conversation just a few days ago can now speak in coherent sentences.

He is thirsty. Dr. Pomahac has given her permission to start him on a tiny amount of water, although in this case the water is first mixed with a thickening agent. Jim's throat is recovering from the trauma of the surgery and the trachea tube is still in place, so swallowing is a bit of an effort. They don't want any liquid to slip past Maki's weak epiglottis and into his lungs. That could lead to pneumonia. Thickening the liquid first helps keep everything headed in the right direction, down the esophagus.

Thin liquid can squeeze past even the tiniest crack and into his
windpipe.

"I'd like real water, not this stuff," he says to her, sucking on
something resembling wallpaper paste.

"You're doing really well, Mr. Maki. Try to be patient."

After his drink Lorrie brushes Jim's teeth for him again. Ensuring that his mouth is pristine and free from excessive germs is
critical to keep infection at bay. He is still on very high doses of
two antirejection drugs, and his immune system is compromised.
But Jim Maki couldn't care less about the germs in his mouth right
now. He has something else in there that is far more interesting.
Teeth! Not dentures, but a full set of real teeth on the top of his
mouth. The donor sure had a lot of fillings. Maybe he didn't brush
enough. Jim will still need a bottom set of dentures to match the
real teeth on top. Dr. Suzuki will make them once the swelling goes
down.

11 a.m. Helfgot residence.
Susan's brother, John, a lean southerner who restores antebel-
lum mansions, has been in Boston ever since Joseph's death last
week. He's in no hurry to leave. The homes he works on aren't
going anywhere, and he likes being here for his older sister. He
sits at the kitchen table, sipping a beer and fielding the flurry of
phone calls that started early that morning.

"No, she can't come to the phone right now. Can I take a
message?"

"No, she isn't doing any interviews."

"No, I don't know if she will be doing any interviews."

"No, she doesn't."

"No, she won't."

"No, she can't."

Every now and then he gets up and ambles outside in his

black boots and skinny jeans, his long hair tucked under a ban-
dana like Steven Van Zandt from Springsteen's band. He sticks
his head into car windows and suggests, not too politely, that
they might want to consider moving on.

Sometimes the cars leave. Sometimes they don't.

chapter twenty-five

Tuesday, May 5, 2009. Helfgot residence.

On the TV a small woman is facing a room full of reporters. Her face is unnaturally swollen. With a slightly impaired voice she tells the cameras, "Well, I guess I'm the one you came to see today. My name is Connie."

Susan's hand stops, her paring knife halfway into a potato as she watches the nightly news. *The woman from Cleveland!* She shoots a look over at Jacob, who sits at the kitchen table pushing around his homework. She tries to gauge his reaction to seeing a woman on television who received a face transplant. He doesn't seem interested. "Is this upsetting you, sweetie? I'll turn it off."

He shrugs a no in that trying-to-be-cool way adolescents have. "Mom, can you write a note to my math teacher?"

Susan has written a lot of notes to teachers in the past few years. The kids' concern for their sick father has taken precedence over the names of the major rivers of the world or all the state capitals. But Joseph has been dead for almost a month. Maybe it's time to get back on track.

Back on track? What goddamn track would that be?

"Sweetie, it's time to get back on track." She tries out the line in a voice she hopes will sound right, somewhere between *I know it's really hard because you miss your dad so much you think you will burst and how could a mother possibly expect her kid to do homework at a time like this?* and *Do your damn homework, Jacob.*

His eyes brim with tears. "I miss Daddy."

"I'm sorry, Jakey, did I sound harsh?"

"Mom, I don't know how to do this part. I wasn't there when the teacher went over it."

His brother comes through the door. "Mom, what's for dinner?"

"Ben, please help Jacob with his math." Susan shuts off the television. They have moved to another story. She goes back to the potatoes.

Two days later she watches Diane Sawyer interview Connie Culp on *Good Morning America*. Culp tells Sawyer that her injuries occurred when her husband shot her in the face. Sawyer wipes the woman's tears with a tissue. Connie Culp says she can't feel the tears. *Good Lord.* They cut to Connie in her bathroom putting on mascara. *It must be easier to be a woman than a man if you have had a face transplant. You can do a lot with makeup. Susan, stop. You're being ridiculous.*

The phone rings and she turns down the sound. Esther Charves has stayed in touch, and now she's a little concerned because Connie Culp has been in the news for several days.

"Yeah, I'm watching it now."

A few weeks ago, when Susan shared her story with the *Boston Globe*, some members of the press wanted more and camped out at her house for three or four days. Katie Couric's people called, asking for an interview, and so did Diane Sawyer's. And now, less than three weeks later, Connie Culp has gone public. It doesn't feel like a coincidence.

Peter Brown from the Brigham told Susan last week that the

man who received Joseph's face would soon hold a press conference at the hospital, and perhaps she might like to participate.
Would she be willing to read a statement?

Before he asked Susan, Brown had a long conversation with
the organ bank team. "Mrs. Helfgot has already expressed an interest in meeting the recipient," they told him. "And Mr. Maki
knows her identity because of the newspaper story. All that's left
is to introduce them." There was another legal document for
Susan to sign, granting permission for the television documentary crew to film her first meeting with James Maki.

"Mrs. Helfgot, we would like you to meet Mr. Maki a day or
so before the press conference. That way, you will have a chance
to get to know one another."

She is eager to meet the man who received her husband's
face. The country needs more organ donors, and Susan wouldn't
be in this miserable place if Joseph hadn't had to wait so long for
a heart. The press conference will be a good platform for talking
about organ donation.

Susan wonders about the man she will soon meet. She still
knows nothing about him. What does he look like? In her mind
she keeps seeing a male version of Connie Culp, but with Joseph's prominent nose. When she describes the images to Esther
they both burst out laughing. "Susan, he can't possibly look anything like that woman. His surgery was different, for starters."

Esther is right. Although both recipients were given new
faces, their wounds occurred in very different ways and their surgeries were completely different. Unlike heart or lung or kidney
transplants, every facial transplant is unique.

"Don't worry," Esther tells her. "You'll see what he looks like
soon enough." Peter Brown has arranged for Susan to come in
to look at pictures of Mr. Maki before she meets him. Esther and
some of her colleagues from the organ bank will be there to support her. They want her to get used to Jim's reconstructed face

before they meet. No one, including Susan, knows how she will react.

They talk about Connie Culp. "I still can't believe what she went through," says Susan. "Twenty-seven operations, and the whole time knowing it's because your husband shot you in the face." *My God. The lives some people are forced to endure.* Susan thinks of her own marriage and how much Joseph loved her.

Esther asks how the kids are coping. "And you, Sue?" Esther is one of the very few people who really understand. She knows what families go through when they must delay the funeral until the organs are removed, how they think about their loved one being worked on by surgeons even after death. It is brutal for these families, but it brings purpose to the grief they must endure, which is often much worse in the days after the funeral.

"How am I doing? I don't even know. At the moment I need something to wear to this thing in L.A." Susan and her two boys will soon fly to California for a memorial planned by Joseph's company. Emily and Jonathan will meet them there. They will all be together again, and although it is for a sad occasion, Susan is looking forward to having everyone back together, even if only for a day or two.

"You'd look good in a sack," Esther says.

"You're the best, Esther. Bye."

"Freckles!" she calls out. The dog jumps up and down wildly. "Sit." She clips on his lead and steps outside into the cool spring morning. A man needed a face, and he got one. It had seemed so simple at the time.

May 10, 2009. Trauma and Burn Unit.

"Hi, Jim." John Maki hasn't come east yet to visit his brother, but he saw the press conference with all the doctors and he read the story by Kay Lazar. "How are you doing?"

"I'm all right."

It's the first time in four years that Jim has been able to have a real phone conversation with his brother. John is thrilled. For a long time the only way he knew how Jim was getting on was through sporadic conversations with Cynthia. Sometimes he spoke with Jessica, but she was mostly off at college, mercifully away from the horror that has been her father's life since the accident.

"So, how do you look?"

"A lot better than I did before." John hears him trying to chuckle.

"When are you getting out of the hospital?"

"Pretty soon. I'm going to be on TV. They're having a press conference for me. I'm writing a speech."

He's writing a speech? John never thought he'd hear anything like that coming from Jim. Writing a speech! Well, why not? He's always had a way with words. For years Jim carried around a thick packet of white note cards bound with a rubber band. He told people he was writing a dictionary, one that everyone could use, whatever their education. His plan was to take complicated words and explain them in a simple way. He used to read the *New York Times* editorial pages. Even after Vietnam, when he was in tough shape, he still read the paper every day. He'd underline words he didn't know, and he'd make a new card for each one.

He also used the cards to hustle money for drugs. He'd ask people to think of a difficult word and bet them he could spell it. Everyone knows drug addicts aren't that smart. But Jim could spell almost any word, and he could tell you what it meant too. That brother of his was really something.

"Are you in any pain?"

"I'm starting to feel something here and there, like after you go to the dentist." Every few days they remove a bit of skin under his chin with a tool that looks like a miniature apple corer. They also take some from the new skin they have attached between the thumb and index finger on his right hand. It's beginning to tug a bit in an uncomfortable way. It's good that he is getting back some nerve sensation, but that doesn't make the procedure any more pleasant.

"You'll have to come out to Seattle and visit me when you get stronger."

"Yeah, I'd like that."

In the early 1990s Jim moved in with John for a while. After his wife died, Professor Maki gave John some money to remodel the space above his garage into a bed and bath. His plan was to come out to Seattle every now and then and spend time with John. Jim helped with the remodeling while John went off to work. Before long Jim hired a plumber, an electrician, and a carpenter. He met them all playing poker. John was a bridge player. Both brothers loved cards and were good players. Jim gambled. John entered tournaments.

John had to admit that Jim hired good workmen. And when Jim was around, the house was often full of people. John led a quiet life, but Jim was gregarious, and people tended to follow him everywhere he went. He could walk into a room and the mood would change. "Do you know that a baleen whale weighs . . ." And Jim would rattle off some obscure fact and get everyone going.

John thinks maybe he should send his brother some new clothes, but he'll wait until Jim goes home. He still remembers the clothes that were stolen at the rehab hospital. That was a tough place, filled with people suffering from severe brain injuries. He thought Jim would never make it out of there. John crossed the country five times to make sure they were taking good care of his brother. Now they're talking on the phone as if nothing had happened. Life can really surprise you.

In spite of everything that Jim has put the family through,

John still loves his brother. "I'm going to be looking for you on TV. I'll come east soon to visit."

When he hangs up, John hopes he and Jim can spend time together again. They had fun when Jim came to Seattle to help fix up the house. But drugs put a wedge between them, and one day Jim drifted on. Maybe now they'll have the relationship that has eluded them their entire adult lives. It would have made their parents happy.

John pictures his mother, a smile framing her soft face. It's a pity she isn't here to see the change that has come over Jim.

Trauma and Burn Unit.
Jim slides out the mirror in his tray table and inspects his face. The scar lines running over his cheeks are beginning to lighten a little. He gently runs a finger over the new skin but feels nothing. Dr. Pomahac has told him it will take time for the nerves to start working. But his fingertips feel something. He traces a zig-zag across his cheek. It's definitely scratchy. He peers more closely into the mirror and rubs his chin.

"Lorrie!" She runs in.

"What's wrong?"

"I think I'm growing a beard!"

"Of course you are. You have a real face now."

"I could never grow a beard before."

"Since the accident?"

"No, I never could."

He takes a long look at the stubble that has sprouted across his entire lower face. "I really have a beard."

The other day he received a visit from Isabelle Dinoire, the world's first face transplant recipient. She came to see him during her trip to the United States. She was a small, shy woman who spoke in French through an interpreter. But she didn't really need to say anything. Her face said it all.

A month after her transplant Dinoire had gone through a bout of rejection. She was alone in the hospital, feeling grateful for her new face, but not seeing many people and wondering when her life would ever resume. It was a difficult time for her. Now her scar lines are virtually invisible, and here she is, traveling overseas.

Jim knows that his face will improve with time, but the waiting is hard. "But you will do better and better," Dinoire reassured him, a smile on her lips. He believes her. She knows.

chapter twenty-six

Susan is bored, and the long flight to Los Angeles is wearing on her frayed nerves. From Boston it takes longer to fly to L.A. than to London. She has finished a mystery novel and is looking over their hotel reservation, mostly for the pleasure of seeing that welcoming, soothing phrase one more time: *early check-in.*

She is exhausted and can't wait to get to the pool. She isn't looking forward to Friday's memorial program. When she gets home, she will meet James Maki, followed by the press conference. She needs to pace herself, just as if she were running.

She plans to run tomorrow morning, taking her old route from the hotel north to Chautauqua and over to the next cliff, past their first house on Via de la Paz.

She and her father used to walk down the street to the cliff's edge, talking about the kids and Joseph's business. Her father always said that he was thinking of renting a small apartment next time, so he and her mother could stay longer. He and Susan would look for dolphins as they stood under the swaying eucalyptus trees, whose heady scent mixed with the smell of the sea. But then, one day, there was no next time. He was gone in less than a year from a deadly brain tumor.

The cliff dissolves and the face of James Maki appears under her eyelids. Not his new face, but the one that was horribly disfigured. She can't get that haunting image out of her head. Last week she and Esther sat at a computer screen in Peter Brown's office. *Boston Med* filmed it. Susan was wearing a wire, and the crew asked her a lot of questions. Now she can't remember a word she said. Who remembers anything at a time like that?

One week earlier. Office of Public Affairs.
Susan gazes at the first picture and then quickly turns away from the camera. She doesn't want them to film her reaction. Her husband's nose is now on a man with brown eyes and dark hair. And it really *is* Joseph's nose on Maki. Wait, *Maki?* His name is on the picture. *He's Japanese? A Jewish nose on a Japanese face: Does that even work? I guess it has to.* She forces herself to focus on the details in the photograph. His face is huge, swollen and puffy, just like Connie Culp's. One eye is sewn shut. She can see the black stitches.

"Here's what the gentleman looked like before," says Peter Brown as he clicks on another file. Susan sees a shell of a face with a large crater in the middle. *Holy Jesus.*

"And this is what he looked like when they brought him in." He shows her several black-and-white pencil drawings.

"There aren't any photos from the time of the accident?"

"Dr. Pomahac doesn't want anyone to see them. They're too gruesome."

May 13, 2009. Los Angeles.
As the plane begins its descent toward the Los Angeles basin, the Salton Sea glistens like aluminum foil in the late morning sun.

Susan can't stop thinking about meeting James Maki. She still isn't sure it's the right thing to do. She appreciates the drama of their meeting, but what about the four Helfgot children? Right after the *Globe* story, when the press was hanging around, Jacob and Ben had to leave the house by a side door. And how will Emily and Jonathan feel about the attention that may come their way? There aren't many situations in life with no real precedent, but this surely is one of them.

Now they seem to be over Joshua Tree National Park, a bit farther south than usual. When she and Joseph lived in L.A., she sometimes spent the night there, lugging her telescope and pup tent into the desert with other stargazers, braving the nightly temperature plunge. As dawn approached they would hunker down to sleep and awaken in the bright sun, broiling like lobsters in their sleeping bags. Joseph never understood her passion for twinkling dots in a dark sky, but then, she never understood his fondness for sea urchin. There are certain things that couples just can't share.

As she checks her seat belt, an image of the astronomer Tycho Brahe pops into her head. It's been like this since Joseph died, random thoughts coming and going more often than usual. She studied Brahe in school. His student, Johannes Kepler, used Brahe's precise observations to come up with the laws of planetary motion in the early seventeenth century. It has always irked Susan that Kepler got most of the credit when it was Brahe who performed the painstaking observations. And he made them without a telescope, which wasn't invented until eight years after his death.

That's the way it works in science. It took three dozen people in two operating rooms to perform the face transplant, but Dr. Pomahac is getting all the attention. Although he deserves much of the credit, he is the first to acknowledge that without Dr. Pribaz and Dr. Eriksson, it would never have happened.

But that's not the real reason Susan is thinking about Brahe. She has often seen his portrait in her astronomy books, the handsome young Dane sporting an outlandish neck ruffle with a singular oddity jumping off the page. Tycho Brahe had no nose, the result of a duel when he was a student. Fortunately for Kepler, and for science, Brahe survived the incident, sporting various artificial noses made of silver and gold, and sometimes copper. (Copper was lighter, but it turned his skin green.) To keep his prosthetic nose in place, he kept a small jar of glue in his pocket. Medicine has come a long way.

"We're stopping at In-N-Out, right, Mom?" Ben asks as the plane turns north.

"Benny, can you see Grandma Rachel's grave?" He is in a window seat on the right side of the plane.

"I can't see anything," he says. "It's too murky." A thick layer of ocean fog envelops the plane as they approach the Pacific.

"So we're stopping, right?"

"Let's see what the driver says." Every Helfgot trip to L.A. begins at the In-N-Out Burger next to the airport.

Someone from Marketcast is sending a car for them, but Susan isn't sure if the driver will make an extra stop.

When Joseph was too sick to drive, a man named Cesar took him around to meetings and the three places her husband deemed essential: In-N-Out Burger, Tito's Tacos, and Pink's Hot Dog Stand.

Cesar was kind, decent, and hardworking, carrying Joseph's bags when Joseph was too weak to lift them. He would pick up medicine and seltzer water to quench Joseph's burning thirst from the strong diuretics he had to take. Cesar would meet Susan and the kids when they flew in from Boston and take them anywhere they wanted to go.

Once, when Joseph was still driving, he passed out and flew through a stoplight and up onto a curb, smashing into a parked car. He was lucky he didn't kill anyone, or himself. Dr. Gregg Fonorow, his physician at UCLA Medical Center, patched him up with some additional drugs that worked for a while.

She has no idea how to get in touch with Cesar. The company asked her to send back Joseph's BlackBerry when he died. It was company property, they said, but to Joseph it was his life. She downloaded Cesar's number along with those of some other friends, but Cesar's cell phone number was disconnected when she tried it the other day.

Cesar doesn't know Joseph is dead. He may think we just decided to blow him off for no apparent reason. It's been two years since Joseph was here. And now he will never be here. She remembers the night they dined on a patio rooftop looking out at the Hollywood Hills and made the decision to sell everything and move to L.A. to give Marketcast a fair chance. It was a gamble, and it worked. They got lucky, but only financially. As the plane lands she starts to cry, and she peers out the window so her boys can't see her face. *I wish I could call Cesar.*

"Hi there. I'm Kirk." A not-so-young man with bleached blond hair flashes a wide smile. His overbuffed arms are holding up a sign that says DIVA Limo. Under the logo is written H-E-L-G-O-T. *People always miss the F.* Susan can hear Joseph saying, *No, that's not right. It's H-E-L–F as in Frank–G-O-T.*

"I'll be your driver while you are here."

"Can you take us to In-N-Out?" asks Jacob.

"Sure."

It takes forever to crawl the one block from LAX to In-N-Out Burger. Kirk tells them all about the house he is planning to buy with his dad if they can scrape together the down payment, and how his daughter in high school wants to be an actor, just like him. He's had some parts on daytime soaps and had a good run as an extra on *Baywatch* that turned into a few lines here and there, but the show ended a while ago and things have been slow since then.

Now he's trying his luck in films. From their itinerary, he's assuming the Helfgots are connected to the movie business. On Friday morning he will pick them up from their hotel and take them to the ArcLight Theater. He's trying to figure out if they are on the business side or the talent side. He's guessing the business side.

"Mom, do you want anything?" Ben and Jake have decided the drive-thru line is too long. They are going inside to order.

"I'll have a number three with a chocolate shake."

She watches her boys run into In-N-Out, their first visit without their dad. This time they'll skip the fish taco place before heading over to the hotel. Joseph loved fish tacos, but he was the only one who did.

They always stay at the Miramar in Santa Monica, always in Room 209. The suite has a counter with bar stools, and when the boys were little they would play bartender and make their parents pretend drinks before going off to swim in the hotel pool. Susan can't wait to get there.

Today they are booked in Room 309. There's a problem with the plumbing in 209, but the woman in reservations has assured her that the rooms are identical. "Everything is exactly the same, Mrs. Helfgot, just one floor up." *Exactly the same? If only!*

"I'm sorry, Kirk. What were you saying?"

"I tried out for a small part in *The Mist* . . ."

"Kirk?"

"Yes?"

"Did anyone tell you why we are here?"

"No, ma'am, they don't tell us anything. They're not allowed to."

"My husband passed away a month ago."

"Really? I'm sorry."

"He worked here in L.A., but we live in Boston. We're here for his memorial on Friday."

"Oh, is that the thing at the ArcLight?"

"Yes. So we may not be a very happy bunch. If we seem to be, you know, ignoring you or something, I just wanted you to know."

He turns around to face her. "How would you like me to act?"

"I'm sorry?"

"How do you want me to act?"

"Just be yourself."

Late afternoon. Trauma and Burn Unit.

"Are you sure?" Lorrie MacDonald isn't sure. Not at all, but Dr. Pomahac seems to be.

"C'mon, let's just do it." Jim Maki is eager.

Dr. Pomahac's beeper goes off. "It's okay," he says. "He's ready." And he disappears.

Lorrie stands with her hands on her hips, her head sideways, assessing her patient. She shakes her head. "I don't know."

"Dr. Pomahac says it's okay."

"All right, but if anything goes wrong, it's not my fault."

"Can we just do this?"

Lorrie puts on gloves and gently places a towel over Jim's face. She presses her hands down on either side of his nose, making sure her fingers are covering his scar lines. "Okay, are you ready?"

"Yeah."

"You're sure?"

"I'm sure."

"Okay, then. Go ahead. Slow and gentle. *Real* slow."

Jim takes a deep breath and then slowly releases it through his new nostrils. He gives a final hard push.

"*Slow!* I said *slow!*"

"I think it worked."

"You did it!"

James Maki has just blown his new nose for the first time.

chapter twenty-seven

Tuesday, May 19, 2009, early morning.
Trauma and Burn Unit.

a t the start of their training, medical students learn how to spot the three classic signs of infection that have been known since medieval times: *calor* (fever), *dolor* (pain), and *rubor* (inflammation). Rejected tissue turns an angry red when the immune system detects foreign cells, treating them exactly like an infection. Large doses of chemicals are released at the site, dilating blood vessels and causing blood and plasma to leak into surrounding tissue, making it swollen and red.

Dr. Stefan Tullius, chief of transplant surgery at the Brigham, stands at James Maki's bedside. What he sees isn't pretty: the allograft has become inflamed. Is it being rejected? The flap from Helfgot is now a bright crimson, right up to the scar line that joins it to Maki's own face. The inflammation signifies something, but what, exactly? Dr. Tullius and his colleagues aren't quite sure. Whatever is going on doesn't seem to fit normal rejection parameters.

The latest biopsy from the sentinel flap on Maki's hand is

clean; there are no signs of rejection. The biopsy from his face is inconclusive. But the skin on his face is definitely inflamed, a classic sign of rejection.

Skin can turn red for several reasons: embarrassment, exercise, even love. Cold and heat have a similar effect. It's not clear that Maki's inflammation indicates rejection, but they can't afford to ignore it.

Lorrie MacDonald, his nurse, says the redness is transitory. It sometimes fades when the doctors leave, or after a biopsy is completed and Mr. Maki has had time to rest. It's peculiar, no question about it. For now they will classify it as mild rejection and treat it with a steroid bolus.

Brookline, Massachusetts.

A videographer walks down the street, past large old homes and a few brick apartment buildings, shooting background of Joseph Helfgot's neighborhood. Large trees buckle the sidewalk, making his steps difficult. But he's getting some good footage and is engrossed in his task.

Brookline High School sits at the end of the block, and the street is alive with traffic. Teenagers, punks and nerds alike, are walking to school in the cool morning air. Skateboarders whiz along and cars crammed with kids smoking cigarettes race down the narrow street. He thinks his producer will be pleased.

He is careful not to film any of the kids' faces. That's a huge no-no. He takes a final long shot and heads back into the house to see if his buddy has finished wiring Mrs. Helfgot for her historic meeting with James Maki.

They stand around the kitchen sink making small talk as they prepare to go to the hospital. Susan slips a last spoon into the dishwasher. In the distance police sirens wail. The sounds get louder until they come to a stop right outside the door.

Susan worries there's been an accident. Maybe a skateboarder has been hit. It's happened before. She peers out the window. The police cars are right there, with lights flashing. *What on earth is going on?*

As she opens the kitchen door a policeman pushes it open from the other side, stepping past her into the house.

"Ma'am," he says in a severe voice. "Do you know these men?" He takes a stern look at them.

"Yes," she says, the word coming out like a question. "What is—"

"Are you *sure?*" he interrupts.

"Yes."

"Somebody called the station about a guy taking pictures of boys going to school." He's staring at the large video camera.

"Oh, I see." She shakes her head. "No. That's not right. They're here about my husband's face." *Joseph's face? What kind of stupid thing is that to say?*

The officer looks at her, trying to process what he has just heard. His partner arrives.

"You know, the face transplant a few weeks ago?"

"Your husband got a face transplant?"

"No, he gave the face. They're making a show about it."

"Which of you guys is her husband?"

"Nobody's my husband."

"We're making a documentary," says the videographer.

"So where is your husband?"

"I told you, he donated the face."

The policemen are confused.

Hoping it will help, Susan adds, "He's dead."

They stare at her. She decides not to say any more.

"We okay here?" the second officer asks his partner. He is surveying the scene outside. A group of kids has spilled off the sidewalk onto the street, blocking traffic. They are curious about the police cars and happy for any reason to be late for school.

"You sure you're okay?" the first policeman asks Susan one last time.

"I'm fine, really."

The moment the door closes, they start laughing. "It's too bad we don't have *that* on camera," one of the men says. "Damn, that was good."

Office of Public Affairs. Brigham and Women's Hospital.
"Before you meet Mr. Maki, he asked me to share something with you."

Susan is spent, but her eyes are alert. She and Peter Brown are alone. Her wire is turned off.

"It's about his past."

"I don't know anything about him, really."

"I know. But the fact that you're meeting with him suggests you are supportive. People will take it to mean you are endorsing not just the transplant, but also the man. At least, that's how it will be perceived at the press conference."

"But I *am* supportive of him." *What is he getting at? He seems uncomfortable.* "I'm not really sure what you're getting at, Peter."

"Mr. Maki has a history of drug use."

"Oh."

A sixty-year-old who has used a lot of drugs. He probably got arrested for drugs. "The organ bank said he was sixty, so I'm not that surprised. You don't reach sixty as an addict without some run-ins with the law, right?"

"Well, it's a bit more than that."

"Oh." *More than that? What else is there that I don't know? Did he murder someone? Please tell me he didn't kill anybody.* "What *exactly* did he do, Peter?"

"Mr. Maki has a criminal record. He has spent time in jail."

He sits back, waiting for her reaction. He clicks his pen once or twice. It sounds awfully loud. He waits some more. Will all hell break loose? He certainly hopes not. It is unbearably quiet in the room as Susan digests the news.

Thank God he hasn't killed anyone. This is so intense. My heart feels like it's going to jump out of my body. What was I thinking when I said yes to all this? Breathe, Susan.

"Oh," she says.

Joseph, are you okay with all this?

Suze, you still would have said yes. You know you would. No one should live without a face. He didn't do anything that bad. He was a drug addict, feeding his habit.

Brown waits awhile, expecting something more. Finally he says, "Are you okay?"

"I guess so. Thank you for telling me. I really appreciate it. It doesn't change anything."

An hour later.

Do I really want to know this man? Yes, I have to talk with him. I have to tell him it's okay, and he shouldn't feel guilty about getting Joseph's face. God, I'm scared.

The organ bank sent Susan a pamphlet about transplantation. There was a section on survivor guilt. As if Susan doesn't know all about survivor guilt. She could have laughed out loud. *Let me introduce you to the Helfgot family.*

Guilt consumed her mother-in-law like a slow-moving cancer. As a young child Rachel watched her mother burn to death, unable to do anything to help. Her husband was forced into a work camp and her children snatched from her in the Warsaw ghetto. She tried to make the men stop, but she was helpless against them. Then she endured day after day in Auschwitz. Will today be the day they take me to the gas? She always made it to the next selection, but millions did not. Mountains of guilt pressed down on her heart. It's a wonder that she never went completely mad. Her past made her eccentricities seem tame.

Rachel passed her guilt on to Joseph, and Susan has often wondered how he avoided becoming crazy. She knows that he would have wanted Jim Maki to move on without guilt. *I have to talk to this man.*

Trauma and Burn Unit.
Jim Maki walks around his room, pacing. He sits down and immediately stands up again. What's taking them so long? All he wants to do is thank the donor's wife the moment she walks through the door.

Finally Susan and the camera crew arrive. There are more people and cameras in the room. Jim Maki sits in a chair, wearing a Brigham baseball cap. A "Precautions" sign is posted on the door. Susan opens a small closet on a nearby wall, pulling out a yellow gown and a pair of gloves. *I wore a yellow gown every day when Joseph was here.* She slips it on in one quick motion.

James Maki stands up. There seems nothing more natural than to just go over to him and shake hands, but it feels contrived. *I feel like I know this man. Blood really is thicker than water.* The handshake instantly dissolves into a hug.

"Thank you," Jim says to her.

"It's okay, really it is. It's what Joseph would have wanted." She is thankful that she has seen the pictures. "You look good, Jim."

"You have children?" he asks Susan.

"Yes." They sit down. She tells him about the four Helfgot kids. It is difficult to speak. There is so much she wants to say. But not here.

"You have a daughter."

"Yes, Jessica. She may be going to South Korea to teach English. She just graduated from college." He is proud.

"How are you feeling?" Susan is already worried about his health. It is important that he stay healthy, that Joseph's life will somehow continue through this man. If Maki should die, her sole consolation will die with him, and the anguish she is barely able to keep at bay will break her in two.

"I feel great."

"I'm really glad." It is hard to have a real conversation. Susan shoots a quick look at Peter Brown, remembering their talk. She sends an almost imperceptible smile in his direction.

He has Joseph's nose. It is his nose. I can tell. They talk a while longer. As she stands to leave, he stands too and offers his arms. As they embrace she suddenly plants a kiss on his cheek. *I just kissed Joseph's face. This is unbelievable. I have to get out of here before I start crying.*

He calls out after her, "See you Thursday for the press conference."

She is halfway down the hall with no idea how she got there.

"Mrs. Helfgot, do you mind if we ask you some questions on camera?" a woman asks her.

"What? Oh, sure." The questions come fast and furious. Later Susan remembers almost nothing from the interview. She hopes that when the show finally airs nothing she said will embarrass her children.

chapter twenty-eight

Thursday, May 21, 2009, late morning.
Brigham and Women's Hospital.

Jim Maki and Susan Helfgot are hidden from view outside the auditorium, where the press conference has already started. Dr. Elof Eriksson is speaking: "One month ago . . ."

One month ago my husband had been buried for eleven days. I am on funeral time. They're keeping real time.

"I want to thank Susan for the gift she and her husband have given to James, the gift of a new face. And I want to thank James for taking that important first step."

God, it's hot. I wonder if Jim thinks it's hot. His face seems flushed.

"He is a pioneer," Dr. Eriksson continues.

Susan has noticed that Jim instinctively touches his finger to his neck whenever he speaks, a habit he developed when he had a trachea tube. When a tracheostomy is performed, air flows through a tube inserted below the vocal cords, making speech impossible. To allow a person to speak, the tube must be capped, forcing air up across the cords. In the 1980s David Muir, a man with muscular dystrophy, invented a small button valve that can be placed over the trachea tube, making speech possible. Until then most patients just covered up the tube with a finger when they had something to say.

Joseph had a bright purple valve. It's still in the night table drawer on his side of the bed. Sometimes it popped off when he coughed, and would skid across the floor, forcing her to crawl around to find it. She would wash it, making sure she dried it completely before giving it back to him. If a drop of water found its way into Joseph's lungs, it could have killed him.

God, I hated that damn thing. Jim probably had one too and just said the hell with it. A lot of people do. They just cover up the hole with their finger and . . . Susan! Get it together. You are about to be on national television and you're thinking about a stupid purple button.

Jim and Susan wish they could just go in and sit down with everyone else and get on with it. Susan is a wreck, and she worries that she'll start to cry when she reads her speech. Nurse Lorrie is with them, trying to help Jim stay relaxed, ready with tissues to mop up the incessant drool from his mouth. Eventually, when Jim's facial nerves start working in his new upper mouth, he will instinctively begin to swallow the twenty or so droplets of saliva his body produces every minute. For now, much of it escapes if he doesn't keep his mouth closed. It's hard for him to control his new mouth. Until last month he didn't even have a mouth.

Susan watches as a slight line of drool starts to drip down the front of Jim's starched blue shirt. Lorrie catches most of it in a tissue. *I hope the spot on his shirt doesn't show up on television.*

It's their turn to speak. Peter Brown leads them into the auditorium. Jim goes first.

"The first part of my life was nothing but trouble." His voice cracks. He thanks the nurses and doctors for giving him another lease on life. "I will be forever grateful."

Now it's Susan's turn. *Don't cry. Whatever you do when you start talking, don't cry.*

She talks about how Joseph worked on movies like *Iron Man* and *Spider-Man*, but that the people who sign up as organ donors are the real superheroes. She asks everyone to visit the website of Donate Life America and register to become a potential organ donor. "Most of all, my thanks go to Jim, who through tenacity and sheer bravery has come so far in such a short time."

As she sits down, Jim looks into her eyes. She smiles at him, and his good, left eye crinkles into feathery lines. Although his face doesn't move, she knows he is smiling back.

They return to Jim's room and sit for a while with Dr. Kim, who asks, "So, how do you think it all went?"

"I'm glad it's over," Susan says.

Jim is tired, his shoulders stooped. "I think we did all right." He takes off his white baseball cap and rubs his scalp.

The film crew is outside. They have been taping throughout the event, but this is a private meeting. Susan has heard from a few scouts who think their story might make an interesting book, and they are trying to figure out their next steps.

Dr. Kim and Susan have already discussed whether a book is a good idea. Dr. Kim thinks it might be helpful for Jim to reflect on his past in a structured way. "The two of you will have a lot of work to do if there's a book, won't you?"

"What else do I have to spend my time on?" Jim is making a joke, and a crinkle forms in the corner of his left eye.

Susan shrugs. "Who knows?" She doesn't expect there will ever be a book. The movie business is filled with options that are never exercised, and she suspects that the world of books works the same way.

"How are you feeling?" she asks Jim.

"Good, I'm going home."

Home? Where? She isn't sure whether it's appropriate to ask him where he lives. *I wish this weren't so awkward.* "You mean now?"

"Yep, today. They're going to take me in a car. I'm ready."

Who wouldn't be? Hospitals are hell. "Do you live far from here?"

"Not too far."

She searches for something to talk about. "How are your daughter's plans to teach overseas coming along?"

"She's in Europe right now, traveling with her girlfriends. She hasn't heard back yet about South Korea." He looks at Dr. Kim. "You're from Korea."

She laughs lightly. "I came here when I was thirteen. And I wasn't traveling alone."

"That must have been a big change. Was it hard?" Susan asks.

"We lived with different relatives for a time."

"I lived in several different places when I was a child," Susan says. "We moved to Boston in my senior year of high school."

"Me too," Jim tells her.

"Really, you moved in your senior year?"

"To Amherst. My dad was a professor at the University of Massachusetts. I grew up in Seattle."

"My husband was a professor when I first met him."

"I heard that."

"Was it hard, moving from Seattle to Amherst?"

"I didn't like it."

"I didn't like moving either," says Susan. "We came here from Fort Lauderdale. My dad bought a house on the South Shore near the water. He thought it would be like Florida, but Massachusetts is nothing like Florida."

"I know what you mean."

Someone from Security escorts Susan to her car, keeping her away from the press. It seems silly, because they know where she lives. The phone is ringing as she walks into the house. It's probably a reporter. *Maybe I should change my number.*

"Hello?"

It's one of Joseph's doctors. Maybe he saw the press conference. "Sue, I have a question for you. Do you know anything about rosacea?"

"It's a rash, isn't it?"

"Yes, on the face. Do you know if Joseph ever had it?"

"I don't think so."

"You're sure?"

"If he had it, I would probably remember. Wouldn't you have known? You guys knew every square inch of his body."

"Well, rosacea's not always easy to diagnose. It can be very mild and go undetected."

"Why are asking me this?"

"I'm not allowed to tell you."

This must be about Jim. His face was flushed at the press conference.

"Do you think Jim Maki might have rosacea?"

He doesn't respond.

"Is it dangerous?"

"Generally, no."

Susan sits down at her computer and logs on to the National Rosacea Society website. The disorder has no definitive cause. It is a vascular condition in which blood vessels dilate, causing severe and sustained blushing. Over time blood vessels may break, and skin tissue, particularly around the nose, may thicken. *Joseph never had any of that. Did he blush a lot? Not really.* She reads on. Flare-ups of rosacea can be tied to emotional stress. "Emotional stress?" she reads again, this time aloud. *Like maybe getting a face transplant and having a press conference? Wouldn't that qualify as stressful?*

• • •

The car chugs up the steep hill away from the river and passes the church. The woman who runs the veterans' home knows that Jim Maki is only a minute away, and she stands on the porch steps, watching for the car.

He eases himself out of the backseat. She smiles at him and they enter the house. "We missed you," she says.

Everything looks pretty much the same. A few of the guys sitting in the dining room look up and tell Jim it's good to see him again. One of them tells him that his face looks great. After a few minutes he heads upstairs to his room. He's happy to be back here with his television and computer. He really missed his computer.

He sits on the bed. The last time he was in this room it was barely spring, a chilly, rainy day. The Red Sox had lost their season opener the night before. Tonight they play the Blue Jays, and Jon Lester is on the mound. In 2006 the Red Sox pitcher was diagnosed with a rare form of non-Hodgkin's lymphoma, but he has made a remarkable recovery. Jim closes his door and turns on the television.

Early July. Department of Plastic Surgery.
Dr. Pomahac fiddles with his PowerPoint presentation. He has been improving it sporadically since the spring, when he put it together to demonstrate the remarkable results they have been able to achieve with a facial allograft. Each time he runs through it he thinks of another nuance or an observation that might be helpful to others who are interested in this groundbreaking reconstructive surgery.

Everyone wants to hear about the transplant, and by now he could probably give this presentation in his sleep. He's become pretty good at speaking with reporters, too. They always ask the same question: "What was the most exciting moment?" He always answers the same way: "When blood began to spread through the allograft, turning it pink. Surreal."

An acrylic model of Maki's skull sits on a table in the corner of his office. Next to it is a model representing the section of Helfgot's face that was used to repair his patient. He clicks away, one slide at a time. A surgeon operates on the notion that there is always room for slight improvements.

James Maki has been doing remarkably well, notwithstanding the rash they have cautiously begun to call rosacea. If that's what it is, it's not a big problem. And if it's a mild form of rejection, they will know soon enough. But the skin guys think it's rosacea, and they turn out to be right.

With Maki doing so well, Dr. Pomahac has begun to allow himself the luxury of focusing on the future. He thinks about the patient who will most likely be their next candidate. He and Julian Pribaz took several patients' files to Brussels last year. The problem is finding more families like the Helfgots.

By the summer of 2009 just two face transplants have been performed in the United States and only a handful of others in the rest of the world. Along with a curious public and a more critical medical community, another constituency has been watching these events with keen interest: the U.S. Department of Defense.

Every war leaves its own stamp of brutality. In Vietnam it was Agent Orange, which led to horrific birth defects and many deaths. In Kosovo it was land mines. In Iraq and Afghanistan it's improvised explosive devices, known as IEDs. First used by the Irish Republican Army in the 1970s, these roadside bombs have been responsible for 40 percent of American military casualties in Iraq; some estimates put the number at 60 percent. Those who are fortunate enough to escape death from IED explosions often lose limbs. The Army Office of the Surgeon General recently reported that well over a thousand American soldiers have undergone limb amputations as a result of IEDs during the Iraq and Afghanistan conflicts.

IEDs are brutal, and sometimes they shear off portions of faces and skulls as well. In 2006 the television journalist Bob Woodruff was almost killed by an exploding IED in Iraq. Surgeons used an acrylic implant to restore a large portion of his skull. He also underwent extensive plastic surgery.

Some wounded warriors have suffered facial injuries so severe they cannot be helped by traditional methods. Two hundred or more American veterans, many living as virtual shut-ins, could benefit from this life-giving surgery. The Department of Defense is determined to find a way to help them. Several successful hand transplants for military amputees have set the stage for the possibility of face transplants. Quietly, on an obscure government website in the spring of 2009, the Department of Defense sent out a request for proposals to conduct clinical studies.

Bo Pomahac filed the Brigham's grant application a few weeks ago. If funding is approved the Brigham's pathology, radiology, immunology, and psychiatry departments will all play a role. Juggling the work will be complicated, but it's the kind of problem that Pomahac and the Brigham would like to have.

He would like to help the soldiers who have given a face for their country. Other than giving one's life, it's hard to imagine a greater sacrifice. He looks up at the model of Jim Maki's skull and imagines several more lined up beside it.

chapter twenty-nine

September 2, 2009. East Madison, New Hampshire.

Susan sits motionless in a tiny kayak, a silent intruder on the shallow pond that is separated from the main part of Purity Lake by a thin strip of lilies. She studies the beaver lodge, searching for signs of life. It seems to have grown higher since last summer. Twigs and small branches jut out in every direction, weaving upward toward a central peak ten feet high.

She scans the surface of the clear, still water, content to wait. Beavers are nocturnal and the day is waning. The fish have forgotten she is here; they jump up along the small molded boat, sucking in stray water bugs. A dragonfly alights on the tip of her yellow paddle. It perches inches from her knee, slowly waving its long wings up and down, basking in a small shaft of sun that splits the boat in half. There is only the sound of the fish and the occasional call from a bird high in the trees.

She closes her eyes and almost falls asleep, lulled by the gentle motion of the boat. For sixteen years she has been coming to this pond, ever since the day they drove back from Mount Washington and Joseph saw a sign for a small resort and decided to have a look. This place is humble and old-fashioned, with Jell-O pudding desserts and a well-used pool table in the basement of the dining hall. Joseph said it reminded him of the Catskills resort where he worked as a teenager. He was fired for serving scalding hot coffee to an important patron who had complained repeatedly that his coffee was always cold. On a bet Joseph had dropped the man's cup into a pot of boiling pasta water and left it there. He fished it out with tongs and cooled the handle with an ice cube before placing the cup on a saucer. He filled it with hot coffee and raced out to the man's table, where he presented it with a flourish. "This cup of coffee is *hot*, Mr. Weinstein. Please be careful."

The other busboys watched in horror as the man screamed in pain. Joseph felt terrible and apologized to everyone, but he was dismissed anyway. It was one of those stories he liked to tell during games of gin rummy under an umbrella at the lake, while his boys begged for quarters for the pool table.

The Helfgots love this place. Even when they lived in L.A. they still made the annual trek back to southern New Hampshire during the final week of the summer, always reserving the same room for the following year before they left. They have forged deep friendships with other guests who come that week. During the year they occasionally come together for a wedding, bar mitzvah, or funeral.

When they were here two summers ago Susan rushed Joseph to the local hospital. The defibrillator in his chest had fired in the middle of the night, not just once but several times, shocking them both awake. The local doctors did what they could. The ambulance driver refused to take Joseph to Boston, claiming he was too unstable. Maybe so, but Susan knew that if her husband went into cardiac arrest, a small rural hospital might not be able to save him. Two years earlier it took a room full of doctors more than three minutes to bring him back when his heart stopped beating, and that heart was even weaker now.

So she wheeled him out to the car and had him lie down in the backseat. She raced toward Boston, praying he wouldn't die before they got there.

Joseph stayed at the Brigham for a month. He went home with a permanent IV that pushed a continual drip of a powerful heart drug called milrinone directly into his heart muscle.

He never went swimming in Purity Lake again. The IV line had to stay completely dry to prevent an infection, and the battery-operated pump could not get wet. A few months later the IV line became infected anyway. By then Joseph's heart was so weak he had to stay in the hospital, waiting for a heart. Dr. Kenneth Baughman came in one day and plopped down on the edge of the bed. Joseph called him Clint, after Clint Eastwood. He was known to be a straight shooter, just like Clint's character in *Dirty Harry*.

"I'm afraid we're not going to get you a heart in time, my friend," he told Joseph. "Your numbers are so bad that I don't think we should wait any longer. Maybe it's time for a VAD."

Seven months later, and missing all the toes on one foot, Joseph left the hospital with a ventricular assist device in his chest. It was already summer again, and the Helfgots returned to the lake, their car filled with medical supplies, including a respirator, because Joseph couldn't breathe at night without artificial support. When a person lies down, his lungs have to work hard to overcome gravity. And Joseph suffered from sleep apnea, which made breathing even harder.

He managed to last for two days and didn't argue when Susan decided it was time to head back. At least he had made it to the lake. The kids stayed on with friends, and everyone waved goodbye as Joseph took a last, longing look at the water. Susan backed the car away from the farmhouse inn and drove away from the mountains. Seven months later he was dead.

This summer she is here with Jacob, having dropped off Ben

at NYU the week before. It's not easy to be at the lake without Joseph and Ben, but not being here at all would be worse.

She spots a bubble trail forming along the surface of the water and sucks in her breath. The bubbles move toward the kayak, coming closer and closer until she finally spots the beaver through the clear water a few feet below the surface. It curls around the boat, investigating for a moment, and then dives deep, pushing its mighty tail against the water, forming small eddies on the surface that dissolve back into glass almost immediately.

The beaver moves to the other side of the kayak and disappears beneath the lily stems. Then it shoots away, leaving another trail of bubbles. Susan follows them until they fade and turns toward the main lake. A loon cries out. She will return here tomorrow. She might see the blue heron.

Guests are scattered on lounge chairs as she nears the wooden bridge. She is glad she can't see the pity in their eyes as they look up from their books. She can almost hear the whispering: "Remember that face transplant? Yeah, did you know Joseph Helfgot was the donor?"

"No, really?"

"Yeah, that's his wife in the kayak."

As she pulls the tiny craft onto the beach, their heads are back in their books. When she plops down on the grass they pretend to notice her for the first time.

The next day Susan and a friend go into town to buy backpacks for their kids at one of the outlet stores. They stop at a gas station where cell phones can get a signal. The scent of pine trees mixes with gasoline as she sits on the curb checking her messages. Jim Maki has called four times in the past couple of days. Why?

"Hey, Jim. What's up?"

She hears the sound of a TV baseball game. "Peter Brown called me. He says Dr. Oz wants to do a show with me."

"Really? That's great, Jim. What do you think about it?"

"They want you and Dr. Pomahac too. You need to call him."

"Okay." She has never watched Dr. Oz.

"Dr. Kim left yesterday."

"I knew she was leaving soon." Susan and Dr. Kim exchange e-mails from time to time. Dr. Kim has decided to spend a year in Germany, working with troops coming off battlefields in Iraq and Afghanistan. She continues to have a close relationship with Jim, who has been relying on her for encouragement and guidance.

Susan almost says, *You're really going to miss her, aren't you?* But something holds her back. She doesn't want to presume to know him that well. They are still almost strangers.

"Well, I guess that's all I had to say."

"Bye, Jim."

In the kayak the next day she replays their conversation. Has Jim ever been to the White Mountains, she wonders, or seen a beaver lodge, or heard a loon's call shatter the silence? Did the Maki family ever spend time in New Hampshire? She stares back at a turtle the size of a fist who is sunning himself on a dead log a few feet away, his long neck pointing in her direction.

Jim has been in the hospital recently, with a small amount of rejection that requires medication. He seems to be doing well, but HIPAA regulations prevent her from knowing anything other than what he chooses to tell. The lack of information about this man sits uncomfortably with her. Maybe it's the abrupt break from the past few years, when she functioned as her husband's personal nurse. No more cleaning sterile VAD wounds, or looking for purple buttons, or taking blood pressures, or arguing over the phone about the delivery of medical supplies to the house, insisting loudly that she can't wait until Monday for sterile gloves because she needs them *now!* In an instant that whole world came to an end.

Is it withdrawal from her daily fix of medical minutiae? Or perhaps a growing concern for the man who wears her husband's face?

"I am having some rejection," Jim told her a few weeks ago.

"They're going to keep me at the Brigham for a few days." That's all he said. Maybe he didn't know the details. Or maybe he's just not a detail kind of guy.

Joseph certainly wasn't. When he and Susan were learning about life on a VAD machine, he didn't want to know too much. And there was so much to know: so many instructions to keep track of, so many warning signs to attend to. She stepped out of his hospital room to call their electrician to make sure the outlet in their bedroom was properly wired for the VAD machine. Later the nurse practitioner told her that the moment she left, Joseph had asked, "Have you seen my wife? Will we be able to have sex with that machine in me? I love my wife."

"Mr. Helfgot, you're having a major medical procedure in an hour. This is a really big deal. Do you have any *other* questions?"

"Yeah. Will I be around for my son's bar mitzvah?"

Maybe James Maki is a bit like that, not one for medical details. *What the hell! It's none of my business.* She laughs at the absurdity of it all, causing the kayak to shake a little, sending ripples across the pond and scattering the nearby fish.

Occasionally she asks Jim, "Are you sure you're okay?" She is itching to ask *What's really going on? Are you running a fever? How high? What meds are they giving you? What do the doctors think? What does Dr. Pomahac say? Is he worried?* But she doesn't probe. After all, she barely knows James Maki. He is still almost a stranger, and maybe always will be. And yet—

Jim is excited about doing a book. He has been telling Susan his life story in fits and starts. He's had a difficult and lonely life, and Susan always feels drained after they have spoken. How do people endure so much?

Two weeks ago Dr. Kim sat with Jim and Susan and talked into a tape recorder. She kidded with him about his reluctance to consider himself brave.

"You weren't ever scared?" she asked.

"No."

"You *are* brave."

"Not really," he replied. But Susan sensed that he was proud of himself. He had signed up to go to Vietnam. James Maki is no coward.

As Dr. Kim spoke passionately about her new challenge in Germany, Susan could feel Jim's anxiety. He seemed lost in thought. This petite woman with long, shiny black hair has a special calm, an inner peace that seems to flow out of her and into the room. Her softly accented Korean voice emerges from a mouth perpetually caught in a small half-smile. Her warm eyes meet you fully, inviting you to share your secrets, but never pushing. Of course Jim will miss her terribly. Who wouldn't? She has guided him through the early weeks of life with a new face and prepared him for his interview with Diane Sawyer in June. She was there for the press conference. But what about *The Dr. Oz Show?* Who will help him through that?

Susan had called Peter Brown from the gas station in New Hampshire. "Dr. Oz used to be a regular on *Oprah*. He has his own daytime show now, with a studio audience. They tape in New York."

"Does Dr. Pomahac want to do it?" She doesn't feel comfortable about it unless he'll be there too.

"I think he does. It's a medical show, and Dr. Oz is a real doctor, a heart surgeon. And it's not for a while."

A syndicated television show in front of a live audience? That's a big deal. Is Jim even strong enough to travel? Do they know how many germs there are on a plane? That can't be good for someone on immunosuppressant medication. Maybe we should take the train. *We? Susan, are you crazy? Let the hospital worry about Jim!*

She starts to shiver. Clouds scuttle across the sky, forming a loose blanket that hides the sun. The water bugs and fish have disappeared. There is no peace here today, and she smacks the yellow paddle hard into the still water, making loud, angry splashes as she glides back to shore.

chapter thirty

October 2009. Helfgot residence.

Jacob shoves textbooks into his overcrowded backpack.

"You're sure you don't mind the monogram?"

"Mom, you've asked me a hundred times. I don't care! I gotta go. Can I have my allowance?"

At the outlet store in New Hampshire, Susan and her friend found monogrammed backpacks on a half-price table. They were identical to the full-price ones, except for the letters. A black bag that Susan liked had the letters WBC faintly visible on the front flap. Susan guessed that Wendy Boyd Chase had asked for black initials on a pink bag, which is why this one was discounted. "Think the kids will notice?" she asked. Half-price meant saving close to forty dollars for each one.

"We could always cut out the threads."

Susan has started shopping the sale racks, which she never used to do. She stood at a window the other day, looking at an expensive pair of Italian shoes, until she finally turned away. When she can't sleep at three in the morning, which is often, the future lurks dark and hazy. Her hands become clammy when she goes online to pay her bills, the same bills in the same amounts that she has been paying for years. All she can see are eight years of college tuition checks. Although there is money to cover those expenses, she feels panic whenever she thinks about it. Jacob heads out the door with Wilhelmina Beatrice Crazowski's backpack slung over his shoulder.

The other day he watched her stuff ten cans of tomato purée into a kitchen cabinet. "Mom, you're starting to act like Dad."

"They were ten for ten dollars."

"Now you're scaring me."

The two of them are still making their way through a case of unsweetened grape syrup that Joseph bought online two years ago. Each bottle makes twenty gallons of juice, and a case makes 240 gallons. Even with a bar mitzvah and a funeral this year, a family can drink only so much grape juice. The case sits on the floor of the pantry under an oversized box of dog food. Other items collect dust in the pantry, thanks to Joseph's shopping habits, including seven packages of tube socks and a box of a hundred Tea Bags of the World that he brought home from Costco. Susan used to cringe whenever he pulled into the driveway after a trip to Costco.

It became more difficult when he had to stop driving. "Can we take a quick run to the grocery store?"

"We went yesterday. There's no more room in the fridge."

"Suze, there's nothing to eat," he would say, scanning the refrigerator with the door wide open, which drove her crazy.

At the supermarket Joseph would work his way through the different brands of olive oil, reading every label.

"Just pick one, please. I need to get home and start dinner."

"This one has chardonnay in it."

"Yes, and it's fourteen dollars!"

"You don't care what you eat."

"I do. That's how we met, remember? The restaurant? Joann introduced us? Which I now am beginning to regret."

"This one has cornichons in it."

"How much?"

"Nine-fifty."

"I'm going out to the car. Let me know when you get to the checkout. And put back one of those crates of clementines. We can't eat two crates. They'll go bad."

"They're on sale."

"Put one back."

"You can make juice."

"I'm going out to the car."

Now I'm buying discarded backpacks and hoarding tomato purée. She tidies up the bathroom. Jacob has left toothpaste goop in the sink and the people from *Dr. Oz* are on their way over.

The segment producer leans against the kitchen counter, sipping a cup of coffee while the camera crew sets up. They are here to film background scenes. B-roll, they call it.

The producer is telling Susan how sorry she is that her husband has died. *You think you're sorry? Imagine how I feel.* Susan thinks this every time people say they are sorry. But she found herself saying the very same thing to a friend whose wife had recently died. What else is there to say?

Susan hesitated when she was asked if they could shoot Jim's interview at her house. This will take their relationship to a new level. He will see the family pictures with Joseph hanging on her walls. He will have lunch at her kitchen table.

She started to say, "I don't know . . ." Then, "Sure, why not?" She can't think of a good reason not to do it, but the question made her uneasy. She sits at her computer in another room while they interview Jim.

"I thought people would act different when I had this surgery, but they still stare at me," he says with regret. He is only six months out from surgery, and Dr. Pomahac and Jim's dentist, Dr. Suzuki, have both assured him that his looks will continue to improve. There will be some dental work and more surgery. Although he looks a lot better than a man without a face, he still has a long way to go.

Mid-November. Back Bay Train Station, Boston.
Susan and Lorrie, Jim's nurse, stand in line for donuts and coffee while waiting for their train to New York, where they will tape *The Dr. Oz Show.* Lorrie is worried about Jim's sugar intake. He asked for six packets for his coffee. "How about just a couple this time, Jim?" she says. But they are in a festive mood and she says it with a light touch. It's hard to be a disciplinarian at a time like this.

"Do you want me to put the sugar in your coffee?" Susan asks. His right arm is useless, and he can't rip open the packets.

"Put in four," he says.

Susan stirs in two packets.

When their train is called, they take the elevator downstairs because he can't negotiate the escalator with his limited sight and lame right arm. As the train approaches, the passengers crowd around. Jim cranes his neck as he hears the train and steps closer to the edge of the platform to get a better look. Lorrie's and Susan's eyes lock in a moment of terror.

"*Jim!*" they scream out in unison.

"What?"

"Don't stand so close to the edge," Lorrie says.

Susan lets out a lungful of air as the train pulls up.

chapter thirty-one

November 18, 2009. New York City.

they leave the train and walk into a sea of people at Penn Station. A greeter from *The Dr. Oz Show* leads them to a car. Jim
climbs into the front and stares out the window. "New York is a
big town," he says to the driver.

"You've been here before, right?" Susan asks. Everybody gets
to New York sooner or later.

"Yeah, lots of times. Forty-second Street is pretty neat."

"You like Broadway shows?"

"Yeah."

He is thinking about Avenue C, but he doesn't want to mention it. Back in the 1970s, right after Vietnam, he would visit
New York often. It was easy to score here. Parts of the Lower East
Side, especially near Thompkins Square Park, had dissolved into
informal anarchy. Buildings were abandoned as competing gangs
vied for control of the tenements that had turned into crack
houses. Few remnants of the old days remained. Eastern European immigrants like the Helfgots, whose store was on Avenue

B, had settled across the river in Brooklyn or over in Queens.

It was easy to score in Times Square too, but there was less chance of getting arrested on the streets of Alphabet City, as Joseph's old neighborhood was called. But it was much more dangerous.

As they crawl along Sixth Avenue they pass a branch of Chase Manhattan Bank. Susan smiles as she remembers the day she took Joseph's mother into the bank's main branch near Wall Street. Rachel had wanted to get her passbook stamped. She never believed the interest was hers until they stamped it in her little blue book. She might have been the last person in New York to still have an actual savings passbook.

They had been to Katz's for lunch. The famous deli has served pastrami and corned beef at the same location for more than a hundred years. Billy Crystal sat there with Meg Ryan in *When Harry Met Sally*, which led to the famous line, "I'll have what she's having."

Twenty-six years earlier. Lower East Side.
"Ma, leave it. Let's go."

"I'm taking it." Rachel pulls paper napkins from a dispenser on the table and tries to wrap her half-eaten pastrami sandwich. Katz's makes a big sandwich, and these napkins aren't up to the task. She closes the messy package and shoves it into her purse.

Joseph drives through the financial district, avoiding FDR Drive on his way to the Brooklyn Battery Tunnel.

"Stop! Stop! I vant to get my book stamped. I need to do dis anyway." Rachel is almost illiterate, but she knows the words *Chase Manhattan Bank*. "I go in here."

"Ma, I can't stop here. I'll take you to the bank in Brooklyn tomorrow."

"Please, I beg you. I vant to go out now. She can go with me." Rachel always refers to Susan as "she." Without waiting for an answer, she opens the passenger-side door and gets out.

"Christ," Joseph says to Susan, who jumps out and follows Rachel into the bank. "I'll drive around."

Rachel and Susan queue up in line. The bank is old and cavernous, with marble everywhere. Businesspeople in expensive suits with impeccable manners wait patiently for the next available teller. Rachel thinks it's taking too long. Each time a teller window opens up, she pushes on the back of the man in front of them. "Hurry up, mister."

He gives them a dirty look the second time she shoves him along. "Rachel, stop it." Susan smiles at the man. He scowls back at her.

As they inch up, Rachel starts searching for her passbook. Various items come out and Susan holds each one: a hairnet, crumpled papers, an empty bottle of celery soda from Katz's.

"Dus is it. I find it." As she pulls out the passbook, a small piece of pastrami falls into the cuff of the man's trousers. He senses something is amiss and looks down. "*What the hell?*"

He furiously shakes his leg in a hokey-pokey motion, trying to dislodge the deli meat. It finally falls off and he kicks it away. A large glob of mustard sticks to the summer-weight wool.

"*Christ*, lady!"

"I'm sorry, Mister." A security guard has noticed the commotion and steps over. He leads Rachel and Susan away from the line and sits them down at a desk.

"But I vas next," Rachel says insistently. She holds out her arm and shows the security guard her numbered tattoo. She always gets a lot of sympathy when she shows the tattoo.

"Rachel, shh!" Susan says. She traces circles next to her head with her finger, and the security guard nods his head.

"Could she just get her book stamped, please?" The security guard stays with them while a manager takes the passbook to a teller window.

"Thank you, you are a very fine man, very fine," Rachel says, pumping the manager's hand before they leave.

"How much is dus?" She hands the book to Susan.

Susan reads the number to her. "Vhat? Such a *bissel?* I think I get more in the Brooklyn bank. Vhere is Yosel?"

"He's driving around the block."

"He should be here."

"He can't park here. He'll be right back."

"Something bad is happened, I know it." She starts screaming Joseph's name. "Yosel!"

"Rachel, please!" Susan yells at her. "He's driving around the block. You're making me crazy, now stop it." Joseph pulls up. "See, here he is. Everything is fine." She pats Rachel on the back and opens the front door for her.

"I am sorry." She looks at Susan. "You need to make a baby for mine son. You are a very fine poi-son."

"How did you make out?" Joseph asks as they climb into the car.

"I vas so scared something bad happened to you."

"Did you get the book stamped?"

He looks at Susan through the rearview mirror. She smiles at him and points a finger to the side of her head, pretending to shoot herself. It's not the first time she has made this gesture in this particular threesome.

"I love you, Suze."

"I know."

November 18, 2009. Rockefeller Center, Manhattan.

"Jim wants sushi for dinner," Lorrie says. "Do you know any places nearby?"

"Let me call my son. I think he may know a place." He does, right around the corner, and Susan feels a little less guilty leaving Lorrie and Jim alone as she jumps into a cab to meet Ben for dinner.

The next morning they meet for coffee before the show. As Jim sips it, some of the coffee dribbles down his chin. After six months his facial nerves have begun to waken, but they aren't yet fully functional. Hundreds of electrical impulses rush from his brain: *Now, purse your lips. Place the rim of the cup on the lower part of your lip; then put the top lip down and take a tiny sip. That's right. You've got it.* The nerve signals run across his face, stimulating the many small muscles that must work together to take one simple sip of coffee.

Susan notices a man at a nearby table staring at Jim. She gets up to grab more napkins and sees that he is still staring when she returns. She shakes off an urge to walk over and say, "What the hell are you looking at?" Instead she stares him down until he becomes uncomfortable and looks away.

A few moments later she takes a quick look to see if the man is still staring. *Are you kidding me? Maybe I should punch him.*

Lorrie has been watching. She says quietly, "A man came up to our table last night while we were eating sushi. He stood there, right in front us, staring."

Susan can't believe it. Maybe he was a doctor and was genuinely curious. "Did he say anything?"

"No."

The Dr. Oz Show tapes next to *Late Night with Jimmy Fallon,* and as Susan sits in makeup she hears the Fallon audience laughing through the thin wall. They are working on her hair. Three other women are also on the show today, and they sit in a row beside Susan. They have until New Year's Eve to lose two dress sizes. Dr. Oz is encouraging them.

"Why are you on the show?" one of them asks Susan. The hairdresser is putting something gluey on the woman's head to give her spikes.

Susan watches, fascinated by the spikes. *So that's how they do it.* "I'm here to talk about face transplants."

"Oh, yeah, I saw your husband in the hallway. He looks pretty good."

"He's not my husband." Susan decides to forgo saying any more. When she tried to explain things to the policeman who barged into her kitchen when the film crew was there, it didn't go all that well.

Jim is in another room getting a quick haircut. Then he changes into the new clothes he bought especially for the show. Peter Brown, who has just arrived with Dr. Pomahac, tells the makeup artist to watch out for Jim's sensitive facial areas.

A woman brings Jim's other clothes into the greenroom. "These are your husband's," she says, handing them to Susan. Lorrie swallows a laugh.

"Sure, thanks, but he's not my husband." She wonders why they think that she, rather than Lorrie, is his wife. Then she realizes that she is still wearing her wedding band.

December 21, 2009. Helfgot residence.
Susan sits down with a cup of hot coffee and brings up the *Boston Globe* online. A picture of Jim and Lorrie sitting together in a hospital room appears on the screen. Susan reads the caption: "Brigham gets $3.4m for face transplants." The grant application went in a while ago, and everyone has been keeping their fingers crossed. Now Pomahac can proceed with more face transplants. Jim told Susan the other day about a man who was in the next room when he got his transplant. He had been shot in the face. He was hoping to get a new face, and maybe now he will.

The hospital will receive funding from the U.S. military to pay for several more face transplants for veterans, and perhaps a few civilians too, who have catastrophic injuries and need this life-giving surgery.

Susan smiles. *Merry Christmas, Dr. Pomahac.*

a personal plea

more than 100,000 Americans are currently waiting for a phone call informing them that an organ they urgently need has become available. Unfortunately, the majority of these people will die before that call ever comes.

Most of them, close to 80 percent, are waiting for a kidney. It's not that kidneys fail more often than other organs, but that dialysis buys extra time. Other members of this group are hoping to receive new lungs, a liver, a pancreas, or most urgently, a heart.

Unlike those who wait in vain for an organ donor, most of us have no idea when, or how, we will die. But in certain cases our death can prolong another person's life. If you haven't yet signed an organ donation card, won't you please do so today? Becoming an organ donor could be the final good deed you perform on earth.

As a popular bumper sticker puts it, "Don't take your organs to heaven. Heaven knows we need them here."

For more information on organ donation, please visit:

Donate Life America—www.donatelife.net
United Network for Organ Sharing—www.unos.org
U.S. Government—www.organdonor.gov

—Susan Whitman Helfgot

acknowledgments

i am indebted to all those who helped and encouraged me during the writing of this book, especially Ellen Ball, Ellen Doeren, Julie Leitman, Julienne Martone, Paula Silver, and Sharon Waxman. My family, especially my sister Maryann Brink, and my friends around the country have been a constant source of support.

I will never forget the medical caregivers at Brigham and Women's Hospital who fought valiantly to save my husband's life. Among many others, I am grateful to Carol Flavell, Leslie Griffin, Lisa Kelley, Kevin McWha, Kristin Morrissey, and Drs. Lynne Werner-Stevenson, James Rawn, Greg Couper, and Gerald Weinhouse.

Esther Charves saves lives every day. She and the many others who work for organ banks around the world are like angels among us.

For their comments on sections of this book, I thank Jennifer Roecklein-Canfield, Stefan Tullius, Julian Pribaz, Bohdan Pomahac, Andrew Selwyn, Marcelo Suzuki, Christine Kim, Kay Lazar, and Sean Fitzpatrick at the New England Organ Bank. John Maki Jr. offered important comments and suggestions. I now know why authors often add a line at this point absolving their early readers of responsibility for any errors in the book. I feel the same way.

Peter Brown has been a thoughtful guide throughout. I am grateful to Lisa Quinn for arranging countless meetings and telephone conversations with hospital staff.

Many other people generously shared their memories and perspectives. I especially thank Joseph E. Murray, Elof Eriksson, Francis Delmonico, Anne Fulhbrigge, Hanka Pomahac, Lorraine MacDonald, Pamela Albert, Kristina Andrzejewski, Christopher Curran, Tenaya Wallace, Joan Ganon, William Doeren, Michael Schwartz, Lucy Wollin, Frank Stryjewski, Katherine Mitchell, Terence Wrong, Carl Hansen, Denise Batchelder, Gregory Ferland, David Lapidus, Jay Hardiman, and Cynthia Maki.

I am indebted to my agent, Ike Williams, who encouraged me to keep writing and persuaded William Novak to read an early proposal, and to Hope Denekamp and Katherine Flynn, who helped the book along in countless ways. Johanna Ehrmann provided excellent advice when I was starting out. I am grateful to Priscilla Painton, my editor, and her colleagues at Simon & Schuster: Victoria Meyer, Danielle Lynn, Mara Lurie, and Michael Szczerban. I am indebted to Judith Hoover, a superb copy editor. Special thanks to Cathy Saypol and Johanna Ramos-Boyer.

Two wonderful friends were lost during the writing of this book. I remember Kenneth Baughman and Brenda Selwyn with deep affection.

This book could not have been written without Bill Novak. In the most sorrowful year of my life I experienced many moments of joy and laughter in his company as he guided each page toward home. I am humbled and filled with gratitude for his friendship and generosity.

To my family—Emily, Jon, Ben, Jacob, Bobby, and Pam—thank you for your bravery.

I applaud Jim Maki's fierce tenacity and his courage in sharing his story. I am honored to call him my friend.

I am indebted to the late John Maki Sr., whose privately published memoir, *Voyage Through the Twentieth Century*, was most helpful.

Some events described in this book happened long ago, and with the guidance of those involved I have re-created a few of them.

Other events took place more recently, often under extreme circumstances. Invariably there are tiny spaces in the collective memory of those who shared a moment of history on a rainy spring night in Boston. I have tried to fill those spaces honorably.

—S. W. H.

Printed in the United States
By Bookmasters